PLASMA
THE FOURTH STATE OF MATTER

PLASMA
THE FOURTH STATE OF MATTER

D. A. Frank-Kamenetskii
Kurchatov Institute of Atomic Energy
Academy of Sciences of the USSR, Moscow

Translated from Russian by
Joseph Norwood, Jr.
Assistant Professor of Physics
University of Miami, Coral Gables, Florida

℗ PLENUM PRESS • NEW YORK • 1972

David Al'bertovich Frank-Kamenetskii, a mining engineer by training, following postgraduate study in 1935-1938, worked for several years in the Institute of Chemical Physics under the direction of Academician N. N. Semenov. His investigations during this period were devoted to the macroscopic kinetics of chemical reactions and to the theory of combustion. Subsequently his scientific interests moved to the field of astrophysics, and he published a number of papers on the theory of the internal structure of stars and the formation of the chemical elements. In recent years, he taught and carried out research work on various aspects of plasma physics. He died on July 2, 1970

The original Russian text was published by Atomizdat in Moscow in 1968. The English translation is published under an agreement with Mezhdunarodnaya Kniga, the Soviet book export agency.

Давид Альбертович Франк-Каменецкий
ПЛАЗМА: ЧЕТВЕРТОЕ СОСТОЯНИЕ ВЕЩЕСТВА
PLAZMA: CHETVERTOE SOSTOYANIE VESHCHESTVA

Library of Congress Catalog Card Number 71-165695
ISBN 0-306-30523-2

© 1972 Plenum Press, New York
A Division of Plenum Publishing Corporation
227 West 17th Street, New York, N. Y. 10011

All rights reserved

No part of this publication may be reproduced in any form without written permission from the publisher

Printed in the United States of America

Foreword

The idea for this book originated with the late Igor Vasil'evich Kurchatov. He suggested to the author the need for a comprehensive presentation of the fundamental ideas of plasma physics without complicated mathematics. This task has not been an easy one. In order to clarify the physical nature of plasma phenomena without recourse to intricate mathematical expressions it is necessary to think problems through very carefully. Thus, the book did not come into being by inspiration, but required a considerable effort.

The aim of the book is to provide a beginning reader with an elementary knowledge of plasma physics. The book is primarily written for engineers and technicians; however, we have also tried to make it intelligible to the reader whose knowledge of physics is at the advanced-freshman level. To understand the book it is also necessary to have a working knowledge of electricity and magnetism of the kind available in present-day programs in junior colleges.

This book is not intended for light reading. It is designed for the reader for whom plasma physics will be a continuing interest. We have confidence that such a reader will want to broaden his knowledge by consulting more specialized literature. Thus, we not only include simple expressions but also special important terms.

Every scientific discipline has its own language. Before undertaking a journey we must acquire a knowledge not only of the geography, but also of the language of the country we intend to visit. It is equally important in the journey to new regions of knowledge that some familiarity with its special language be acquired. Thus, every new term is emphasized (by e x p a n s i o n) when it first appears.

In the American edition of this book we will use the rationalized MKS system of units, which has become increasingly popular in science and engineering in recent years.

The physics of plasmas is a field in which knowledge is expanding rapidly. The size of this book does not allow us to cover everything we should. To mention the work of only some of the workers in this field would be an injustice to the others. Thus, we have decided not to give references or names, except where the names appear in scientific terminology.

The growing science of plasmas excites lively interest in many people with various levels of training. We hope, therefore, that the book will be useful to a broad range of readers.

The author expresses his thanks to friends and colleagues such as S. I. Braginskii, A. A. Vedenov, E. P. Velikhov, V. P. Demidov, E. K. Zavoiskii, B. B. Kadomtsev, I. A. Kovan, V. I. Kogan, M. A. Leontovich, V. I. Patrushev, L. I. Rudakov, V. D. Rusanov, R. Z. Sagdeev, and V. D. Shafranov for their constant interest and many discussions covering the entire range of plasma physics. T. D. Kuznetsov gave invaluable help with the illustrations.

Contents

Introduction	1
Production of a Plasma	8
Plasma Diagnostics	11
Quasi Neutrality and Charge Separation	18
Polarization of a Plasma	22
Gas Discharges	24
Plasma Thermodynamics	25
Elementary Processes	35
Plasmas and Radiation	36
Equilibrium and the Stationary Ionization States	40
Plasmas as Conducting Fluids	44
Field Diffusion and Plasma Diffusion	50
Applications of the Conducting-Fluid Model	51
Toroidal Plasma Traps	53
Electromagnetic Pumping and Plasma Acceleration	55
Magnetohydrodynamic Flow	57
The Two-Fluid Model	59
Plasma Conductivity in a Magnetic Field	64
Plasma as an Ensemble of Independent Particles	66
Drift Motion	68
Electric Drift	71
Conservation of Magnetic Moment	74
The Adiabatic Traps	75
Drift in an Inhomogeneous Field	79
Polarization Drift	83
Rotating Plasmas	85

The Magnetization Current	86
The Quasi-hydrodynamic Approximation	89
Hydromagnetic Plasma Instabilities	91
Pinch Instability	95
Stabilization by Frozen-in Magnetic Fields	96
Interchange or Flute Instabilities	98
Diffusion of Opposed Fields	100
Oscillations and Waves in Plasmas	104
Electrostatic Plasma Oscillations	108
Electrostatic Oscillations with Ions	110
Plasma Oscillations in a Magnetic Field	113
Dispersion near the Cyclotron Frequency	116
Oblique Waves and General Classification of Oscillations	119
Propagation of Radio Waves through a Plasma	120
Plasma Resonators and Waveguides	126
Excitation and Damping of Oscillations	130
Shock Waves in Plasmas	134
Random Processes	135
The Drunkard's Walk	138
The Mean Free Path and the Collision Cross Section	139
Collisions with Neutral Particles	141
Coulomb Collisions	144
The Establishment of Thermal Equilibrium	148
Transport Processes in a Magnetic Field	150
Ambipolar Diffusion	152
A Recent Plasma Experiment	154
Index	157

Plasma — The Fourth State of Matter

Introduction

We have all learned that matter appears in three states: solid, liquid, and gaseous. But in recent years more and more attention has been directed to the properties of matter in a fourth and unique state, which we call plasma. The higher the temperature, the more freedom the constituent particles of the material experience. In solid bodies the atoms and molecules are subject to strict discipline and are constrained to rigid order. In a liquid they can move, but their freedom is limited. In a gas, molecules or atoms move freely; inside the atoms the electrons perform a harmonic dance over their orbits, according to the laws of quantum mechanics. In a plasma, however, the electrons are liberated from the atoms and acquire complete freedom of motion. With the loss of some of their electrons, atoms and molecules acquire a positive electric charge; they are then called ions. Thus, a plasma is a gas consisting of positively and negatively charged particles in such proportions that the total charge is equal to zero. Freely moving electrons can transport electric current; in other words, a plasma is a conducting gas.

At the present time, in electrical technology solid metallic materials are used as electrical conductors. A metal also contains free electrons. They are liberated by forces associated with the high density. In a metal the atoms are so compressed that their electron shells are "fractured." In a plasma the electrons are separated from each other by forces produced by the fast motion of hot particles, the action of light, or an electrical discharge.

The novel properties of plasmas give some basis for expecting new technological applications, both as electrical conductors and as high-temperature media. In electrical applications plasma would have obvious advantage over metals in that it is a thousand, if not a million times lighter.

Plasma physics has become a science only recently, although plasmas have been known to man since early times. Lightning and the aurora borealis, or northern lights, represent familiar examples. Everyone who closes an electrical switch will also appreciate plasmas. The sparks which jump between the contacts represent the plasma associated with the electrical discharge in air.
In the evening in the streets of any large city, one sees neon signs: these too are plasmas, although this fact is not always appreciated. Any material heated to a sufficiently high temperature goes into the plasma state. The transition takes place more easily with alkali-metal vapors such as sodium, potassium and particularly, the heaviest of these, cesium. An ordinary flame exhibits some electrical conductivity, being weakly ionized; it, too, is a plasma. This conductivity is due to a slight impurity of sodium, which can be recognized by the yellow light. Temperatures of several tens of thousands of degrees are necessary for total ionization of a gas.

Under ordinary terrestrial conditions the plasma state of matter is quite rare and unusual. But in the universe cold solid bodies such as our earth appear to be a rare exception. Most of the matter in the universe is ionized, i.e., it exists in the plasma state. In the stars the ionization is caused by high temperature.
In the tenuous nebulae and interstellar gases the ionization is due to ultraviolet radiation from the stars.

In our own solar system the sun consists entirely of plasma, its mass being three hundred and thirty thousand times larger than the mass of the earth. The upper layers of the earth's atmosphere are ionized by the sun, i.e., they also consist of plasma. These upper layers are called the ionosphere and are responsible for long-range radio communication.

In ancient times it was believed that the world is composed of four elements: earth, water, air, and fire. The earth, water, and air correspond to our solid, liquid, and gaseous states of matter. Plasma corresponds to the fourth element, fire, which appears to be dominant on a cosmic scale.

There is no sharp boundary between plasmas and gases. Plasmas obey the gas laws and, in many respects, behave like gases. Why, then, do we speak of plasma as a fourth state of matter? The new and extraordinary properties appear when the plasma is subject to a strong magnetic field. We will refer to such plasmas as magnetoplasmas.

We have noted that within the atom the electrons perform a harmonic dance whereas in a plasma they exist without any order, like the molecules in a gas. However, it is an important property of plasmas that the particle motion can be ordered. The particles can be forced to move in regular fashion. The agency that forces the free electrons to submit to rigid discipline is the magnetic field. In atoms the electrons and the nuclei are clustered in small groups. In a solid crystal they are found in fixed locations. In a magnetoplasma they move collectively in cohesive groups.

The motion of the particles in an ordinary gas is limited only by the collisions which they undergo with each other or with the walls. The motion of plasma particles, however, can be constrained by the magnetic field. Plasma can be contained by a magnetic wall, pushed by a magnetic piston, or confined in a magnetic trap. In a strong magnetic field the plasma particles are constrained to circle around the lines of force of the magnetic field although they can move freely along the magnetic field. The combination of free motion along the field lines and gyration around the lines results in a helical motion. If the plasma particles are forced to move across the magnetic field they drag the magnetic field along with them. We say that the plasma particles "stick" to the field lines or that the magnetic field is "frozen" to the plasma. But this freezing picture only applies in hot plasmas. In a hot plasma the particles pass each other quickly without much interference. Such a plasma offers almost no resistance to electric currents since its conductivity is very high. In a cold plasma with low conductivity, on the other hand, interactions between particles due to collisions allow the magnetic field to leak through the plasma.

In speaking of a "cold" plasma one must realize that the temperature scale for a plasma is not one which we are accustomed to. The temperature unit is the electron volt, which equals 11,600°C. A plasma at a temperature of ten or a hundred thousand degrees is said by a physicist to be at "only a few electron volts — a cold plasma." In hot plasmas the temperature is at least a few hundred electron volts, i.e., millions of degrees. If the gas had not crossed over into the plasma state it would have been impossible to heat it to such temperatures since it could not have been confined. No solid walls can withstand this temperature and the plasma would have dispersed. Hot plasmas can, however, be confined by a magnetic field. The plasma particles describe helices around the field lines and cannot assume another path so long as they do not collide with other particles. In hot plasmas collisions are very rare, as on a street with a well-organized traffic flow. As long as a breakdown does not take place the particles do not hit the walls. This characteristic of plasma behavior is the basis for various schemes that have been proposed for confinement of hot plasmas by magnetic fields.

Unfortunately, it has been found that effects other than collisions can knock particles from field lines. One such cause for the violation of regular motion of the particles is the collective interaction. Imagine a column of trucks moving in regular order along a highway. If even one truck disturbs the order of the motion the disturbance will grow. Those in the rear of the column run into the front, causing others at the front to disrupt the motion of the trucks further back. Although started at one point, this disorder is gradually propagated along the entire column. This effect, in which a small disturbance causes a universal disorder is called an instability. In plasmas such instabilities are very common.

One of the most exciting and interesting problems of modern physics is that of obtaining and confining hot plasmas. The question is one of heating matter to temperatures such that no solid wall could contain it. Only a magnetic field can contain a hot plasma. It must be an impermeable barrier for the various plasma particle and not allow them to escape to the walls or to transfer their energy to the walls. The main obstacle in the quest for the ideal magnetic trap is the instability problem. If it were not for instabilities the problem of confining plasmas could have been already solved by any one of several schemes.

Visualize a tube filled with plasma. By winding a conductor on this tube we obtain a coil. We pass a current through it. The coil thus becomes an electromagnet and establishes a magnetic field inside the tube. The magnetic field lines are directed along the tube, parallel to its axis. If all of the plasma particles move in regular order they are prevented by the magnetic field from hitting the walls. Along the tube, however, the particles can move freely and can escape through the ends. There are two methods to prevent this loss. One is to bend the tube into a circle. This gives a figure like a doughnut, which is known in geometry as a torus. This is a toroidal magnetic trap. The other method is to establish a stronger magnetic field at the ends of the tube. This field acts as a magnetic cork or stopper. The plasma particles are reflected from the strong-field region, which is therefore called a magnetic mirror.

In a toroidal trap the plasma can be heated easily. The trap is filled with gas and a strong electric field is established in the gas. The gas is heated by the current flowing in it in the same way as the coil in an electric heater or a lamp bulb filament. In this way a gas can be heated to the point where it enters the plasma state. However, as the temperature increases the electrical resistance of the plasma decreases strongly and it can no longer be heated. Other methods have been proposed for obtaining high temperatures; there are more refined heating methods that use a high-frequency discharge or rapid compression by a magnetic piston. It is easy to inject fast ions, previously accelerated by an electric field, into mirror traps. In this way we obtain a hot plasma immediately. However, all these methods for heating are suitable only under certain conditions: namely, the plasma must not come into contact with solid walls. It is impossible to heat a plasma in contact with walls just as it is impossible to boil water in a pot made of ice.

If the plasma particles are not to hit the walls they must move in a regulated manner within the magnetic trap. However, instabilities often occur. The situation is reminiscent of leading children through a traffic crossing: the ions and electrons move around in all directions, strike the walls, and thus waste their energies.

Plasma particles bombarding the walls cause atoms of the wall material to be ejected into the plasma. In this way an instability can lead to contamination of the plasma by impurities. The heavy impurity atoms lose their energy in the form of light and ultraviolet rays and this energy loss increases progressively. The trap becomes filled with the cold products of wall evaporation rather than hot plasma.

If a large column of plasma particles always moves "in step" it could be confined equally well in a toroidal trap or in a mirror device. Actually, the plasma order is destroyed, i.e., instabilities arise. A large army of physicists is engaged in a campaign to fight plasma instabilities. The "plasma pacification program" is the subject of their total effort.

In the interior of the sun plasma compression produces temperatures of the order of 10,000,000°K. At this temperature the atomic nuclei collide with such force that they fuse. This process results in thermonuclear reactions that transform hydrogen into helium and liberate an immense quantity of energy. This energy is radiated by the sun and represents the energy source of the earth. Can we "tame" thermonuclear reactions and force them to serve us here on earth? Even at such high temperatures ordinary hydrogen liberates energy very slowly. It is because of the large amount of hydrogen in the sun and the powerful compression force of gravity that solar hydrogen can be such a powerful source of energy. But heavy isotopes of hydrogen — deuterium and tritium — liberate energy at a rate sufficient for our purposes. If we can create the proper containment conditions for a plasma in a magnetic trap at temperatures of tens of millions of degrees, the problem of thermonuclear energy will be solved. To achieve this goal one item is certainly necessary — this is finding a way to cope with plasma instabilities. This is the single difficulty, but it is so serious that no one can say how close the solution is or which way it lies.

The thermonuclear problem has been a motivation for plasma research. A broad knowledge of the plasma properties has been gained because of this research. But, as always, the growth of a science bears unexpected fruit. Columbus sought to find a way to India and found America. In the history of science things often happen this way. In the course of investigating plasma properties it was learned that a plasma can be accelerated by a magnetic field. Plasma guns have been constructed from which plasma can be ejected with velocities up to 100 km/sec. This is a hundred times faster than a bullet and ten times faster than space rockets. The first guns were constructed in order to inject plasma into a magnetic

INTRODUCTION

trap, but it was subsequently realized that a plasma motor could be constructed using the same principles.

In principle, plasma motors operate like electric motors, with plasma replacing metal as the conductor of electricity. But electric motors can be converted into generators by simple switching. Naturally this idea leads to the idea of a plasma generator. In such a generator the deceleration of a plasma jet in a magnetic field produces an electric current. Is it not reasonable that electrical technology could replace heavy awkward metals with lighter plasma conductors?

At the present time the prospects for using plasma in electronics have not been explored. Plasmas in a magnetic field are capable of a large variety of modes of oscillation. Plasmas can emit radio waves. At the present time random oscillations which we call noise have been observed. But the theory indicates that plasma resonators and waveguides can be constructed which will oscillate at completely determined frequencies.

There is much that is unusual in the behavior of magnetoplasmas. Electric forces arise which are caused by mass motions (known as drifts), and forces of a nonelectric nature can induce currents. The velocity associated with motion and the current are not parallel to the force, but perpendicular to it. The force does not cause the plasma to accelerate but imparts a constant drift velocity. All of these unusual properties of plasmas have yet to be exploited for the benefit of mankind. It is quite possible that even plasma instabilities will be found to have useful applications, for example, as a way of exciting plasma oscillations.

Many plasma phenomena are evident on a colossal scale in deep space. Solar flares are evidently due to the rapid compression of plasma by a magnetic field. In this process the flare is expelled into space as a plasma stream. The magnetic field in deep space captures the plasma ejected by the sun in various kinds of magnetic traps. A plasma trap of this kind is found near the earth; this trap is known as the radiation belt and constitutes a hazard for astronauts. After a solar flare, magnetic storms and auroras are observed on earth and disturb radio communications. A solar flare causes disturbance outside the earth's atmosphere by virtue of the plasma streams and shock waves that propagate in the interplanetary plasma. The propagation of shock waves in interplanetary space is one of the remarkable phenomena associated with magnetoplasmas. Space rockets and artificial satellites constantly confirm the role played by plasma in the universe.

In our opinion mankind is entering into the space age which, to a considerable degree, is also a plasma age. This new stage in the growth of science and technology imposes increasing requirements on the youngest branch of physics, plasma physics.

Production of a Plasma

In order to convert a gas into a plasma state it is necessary to tear away at least some of the electrons from the atoms, thereby converting these atoms into ions. This detachment of electrons from atoms is called ionization. The detailed mechanism of ionization on the atomic-molecular level will be examined below. Here we limit ourselves to a discussion of the general characteristics of this process.

In nature and in the laboratory ionization can be produced by various methods. The most important of these are: a) ionization by heat; b) ionization by radiation; c) ionization by electric discharge.

All substances become ionized if they are heated to sufficiently high temperatures. This process is known as thermal ionization. It is necessary that the temperature be close to the energy of the most weakly bound electron, i.e., almost the lowest ionization energy of the atom or molecule. The ionization energy of an atom depends on the position of the particular element in the periodic table. The most weakly bound electrons are in the monovalent atoms of alkali metals (lithium, sodium, potassium, rubidium, and cesium). In such atoms one of the outer (valence) electrons is in an outer orbit and is easily detached. The most firmly bound electrons are in the inert gases (helium, neon, argon, krypton, and xenon). In these atoms the electrons form a closed shell which it is difficult to disturb. In each of the columns of the periodic table the ionization energy becomes much smaller as one goes down the column toward heavier elements (in heavy elements there are more inner electrons which partially shield the field of the nucleus). Therefore, the easiest element of all to ionize is the heaviest alkali metal, cesium. It is often used in the laboratory and in technology to obtain a thermal plasma.

In the presence of alkali-metal vapors the electrical conductivity of a gas can be significant at temperatures of only 2,000-3,000°C. But in order to obtain a fully ionized plasma by thermal means it is necessary for the temperature to reach several tens of thousand degrees.

Stars consist entirely of thermal plasma. Weakly ionized plasmas with high densities and relatively low temperatures can be obtained thermally through the use of an easily ionizable additive. The electrical conductivity of the plasma is associated primarily with the presence of the alkali-metal additive (e.g., the sodium which gives a flame its yellow color).

Ionization by radiation becomes significant in very tenuous gases; with any significant density collisions between particles are much more important than the action of the radiation. This means that ionization is important in astrophysics; the ultraviolet radiation from hot stars causes ionization in the surrounding gaseous vapors and in the interstellar gas (regions of singly charged hydrogen ions* HII). The radiation of the sun gives rise to ionization in the outer layers of the earth's atmosphere. Attempts to use radiation ionization in technology have not been very successful because typical densities are such that the inverse process of recombination of electrons with ions proceeds very rapidly and leads to a condition of equilibrium.

The most widely used method in the laboratory and in technology for obtaining a plasma is the electrical gas discharge. In nature an example of this phenomenon is seen in lightning; in technology, typical examples would be electric sparks, electric arcs, gaseous flash lamps, and other gas discharge devices. Ionization is a discharge depends on the production of an electron avalanche (Fig. 1).

---

* In spectroscopy we observe the following system of notation: a neutral atom is denoted by the chemical symbol and the Roman numeral I, singly ionized ions by the Roman numeral II, etc.

PRODUCTION OF A PLASMA

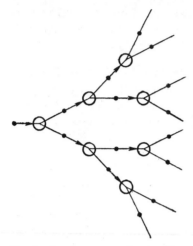

Fig. 1. An electron avalanche (the circles are atoms and the black dots are electrons).

This process is of the same type as a chain reaction in chemistry or multiplication of microbes in an epidemic. In order to produce an avalanche it is necessary that the electric field applied to the gas be large enough so that the energy imparted to the electrons in their mean free paths is sufficient to knock out at least one electron from an atom on impact. These secondary events are sufficient even if there is only a small number of free electrons since these can liberate new electrons after being accelerated by the field. In this way, electron multiplication proceeds in a geometrical progression. Just as a small number of plague microbes can cause a huge epidemic, so a very small number of electrons, produced perhaps by cosmic rays or emitted from a metal surface, can cause ionization of an entire gas and turn it into a plasma.

In addition to these basic methods for producing plasmas, there are many others of less importance. For example, in the search for ways to produce a thermonuclear plasma work is being carried out on injection: ions acquire large velocities in an accelerator and are injected into a magnetic trap; electrons are attracted to the ions from the surrounding medium and together they form a hot plasma.

An unusual method for separating electrons from atoms is the phenomenon of pressure ionization. At very high densities all materials enter into the degenerate state in which the electrons are "squeezed out of" their high energy levels. If the energy of these levels (the so-called Fermi energy) exceeds the ionization energy, then the electron shells "are broken" and the electrons are detached from the atoms. This phenomenon can occur in ultra-dense stars, white dwarfs, and the interior of large hydrogen planets and, according to some authorities, even in the core of the earth. In experiments on compression of matter by converging shock waves an electrical conductivity is observed which can be explained by pressure ionization. However, the necessary densities are so high that the material becomes more like a metal than a plasma. Thus, this effect lies outside the realm of plasma physics.

Plasma Diagnostics

Engineers or experimenters working with ordinary gases have no difficulty in determining the physical properties and state of the gas. Any thermometer or pyrometer can measure the temperature of the gas; a manometer the pressure; a flowmeter, the stream velocity; finally, highly developed chemical and physicochemical methods of gas analysis yield a determination of the chemical composition. Plasmas represent a different situation. Every measurement is a problem. There are many cases in which temperature measurements by different methods on the same plasma yield results which disagree by a factor of ten. Also we find that many experimenters working with plasmas do not even know such basic quantities as the charged particle density with precision.

The reason for this state of affairs is that the determination of the physical characteristics of plasmas, in contrast with gases, does not reduce to a simple measurement. Methods for determining the temperature, density, and composition of plasmas are the subject of an important part of experimental plasma physics which has come to be called diagnostics. To arrive at an opinion concerning the condition of a plasma by readings of instruments is as difficult as trying to establish a diagnosis of an illness by an inspection of a patient. Whereas an ordinary gas responds to the questions of the "doctor" willingly, a plasma only groans like a dumb animal so that the plasma researcher should, perhaps, be compared to a veterinarian.

Many physical phenomena occurring in plasmas can be utilized to provide diagnostics. We will encounter these diagnostic applications throughout this book. At this point we give a general survey of diagnostic methods; the physical principles will be explained in detail later on.

In the absence of a magnetic field the electron density and temperature of a plasma can be determined simultaneously by an electrostatic (Langmuir) probe. The method is based on the polarization of a plasma. A metal probe (Fig. 2) is inserted into the plasma and measures the current to the probe as a function of the probe potential (volt − ampere characteristic, Fig. 3). An important property of plasmas emerges clearly in the probe method: plasmas do not obey Ohm's law. The current is determined simply by the magnitude of the charge which is transported by the positive potential. The current is found to approach a limiting value which is independent of potential. This current limit is called the saturation current and is determined by the charge which is transported by the electrons that strike the surface of the probe in their thermal motion. If the thermal speed of the electrons is known the plasma density can be found from the saturation current. The thermal velocity is calculated from the electron temperature, which is found by inspecting the slope of the volt−ampere characteristic.

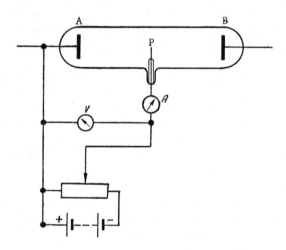

Fig. 2. A method for measurement of electron density and temperature by an electric probe inserted into a gas discharge. A and C are the electrodes (anode and cathode) of the discharge tube in which the gas discharge is excited, P is the electric probe, V and A are the voltmeter and ammeter which give the volt−ampere characteristics. Below is the source of the voltage applied to the probe and a potentiometer which is used to vary the voltage.

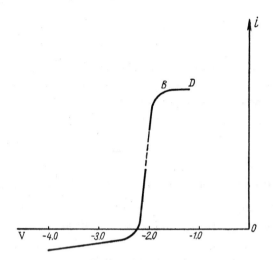

Fig. 3. Volt—ampere characteristic of an electric probe. i is the current to the probe; V is the potential relative to the anode. The segment BD represents the saturation current.

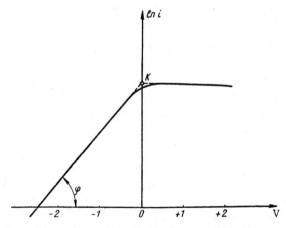

Fig. 4. Volt—ampere characteristic of an electric probe as plotted on a logarithmic scale.

The saturation current is

$$i_s = Sen_e \bar{v}_e,$$

where S is the area of the probe, e is the charge of the electron, n_e is the electron density, and $\bar{v}_e$ is the average velocity of the electrons in one direction. The latter is related to the electron temperature* T_e by the expression

$$\bar{v}_e = \sqrt{\frac{T_e}{2\pi m}},$$

where m is the electron mass and k is the Boltzmann constant.

The electron temperature can be determined with respect to the volt–ampere characteristics in the region in which the probe has a negative potential relative to the plasma. In this region the probe repels the electrons and the surface of the probe can only be reached by those electrons in the Boltzmann distribution which have energies sufficient to overcome the potential difference $V - V_0$, where V is the probe potential and V_0 is the plasma potential. Hence

$$\ln i = \frac{e}{T} V + \text{const.}$$

By plotting the current i as a function of the potential V on a logarithmic scale, we obtain a straight line over a wide range (Fig. 4). The slope of this line allows us to determine the temperature T_e.

The electric probe method has found broad application in the classical physics of gas discharges. In a magnetic field the motion of particles along the field has a completely different character than motion perpendicular to it. Therefore, in the presence of a magnetic field the electric probe is unsuitable for absolute measurements.†

---

* Temperature — in energy units.

† The effect of a magnetic field on an electric probe can be eliminated if the probe dimensions can be made smaller than the cyclotron radius. For probes which capture electrons this cannot be done because the electron cyclotron radius is too small. The ion radius is thousands of times larger. Therefore, absolute measurements can be made with an electric probe in a magnetic field if the ion current is used rather than electron current; we use the ion part of the probe characteristics, which gives the saturation current at large negative potentials.

Readings that follow changes in the electron density can be obtained only if a preliminary calibration of the probe is carried out by an absolute method. In plasmas in a magnetic field the particle density is determined by microwave methods. To understand these methods it is necessary to be acquainted with radio wave propagation in a plasma; this subject will be discussed in an appropriate place.

Optical methods are widely used in plasma diagnostics. The intensity and spectral composition of radiation emitted by a plasma depends on the temperature and, to a lesser degree, on the density of the plasma. Very dense plasmas emit ordinary thermal radiation and the plasma temperature can be determined by optical pyrometry. A more comprehensive knowledge of the temperature, composition, and density of a plasma is obtained by spectroscopic methods. In investigating a plasma its radiation must be resolved into a spectrum. The plasma spectrum is more complicated than the spectrum of a gas or solid body. The gas spectrum consists of isolated, or discrete, lines. The spectrum of a solid body is continuous. A plasma emits discrete lines superimposed on a weaker continuous spectrum.

If there are undissociated molecules in a plasma, in addition to atoms, then in addition to the narrow lines there will be broad molecular bands in the spectrum. A typical plasma spectrum is shown in Fig. 5. From the positions of the characteristic lines in the spectrum, as noted in the picture, it is possible to judge the qualitative chemical composition of the plasma. For example, in the case of a hot plasma in contact with a wall lines quickly appear which are characteristic of the wall material.

At higher temperatures more and more electrons make transitions to higher orbits. The temperature of the plasma can be estimated from the relationship between the intensities of lines emitted from different energy levels. At high temperatures some electrons are detached from complex atoms, leaving multiply charged ions which can be recognized by the new lines they generate in the spectrum. The appearance of lines due to multiply charged ions is evidence of a high plasma temperature and provides an approximate estimate of plasma temperature. The width of the spectral lines in a hot tenuous plasma depends on tempera-

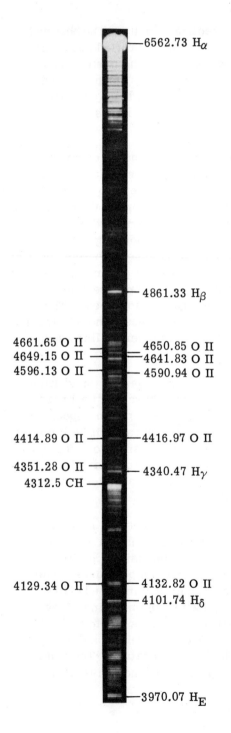

Fig. 5. A typical plasma spectrum. Lines: H, hydrogen atoms; OII, singly charged oxygen ion; CH, the CH radical. The numbers are wavelengths in Ångstroms (Å).

ture; the linewidth in a cold dense plasma depends on the particle density.* The intensity of the continuous spectrum depends more sensitively on the density than on the temperature; if the temperature is known it can be used to determine the electron density. All methods for determining the temperature are useful only when the plasma is in thermodynamic equilibrium. In very tenuous plasmas where such an equilibrium does not exist the idea of a temperature is really not meaningful. Thus, can one speak of a "temperature" in interplanetary space only under certain conditions; the value of the temperature will depend on the definition.

The magnetic field is so closely associated with a plasma that its strength should be included with the physical quantities that characterize a plasma. The measurement of the magnetic field is an important problem in plasma diagnostics. It is useful to employ, for this purpose, a magnetic

---

*In a hot tenuous plasma the linewidth is determined by the Doppler effect; in a cold dense plasma it is determined by the interaction of particles due to their close approach and collisions (collision or Stark broadening).

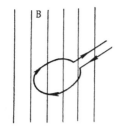 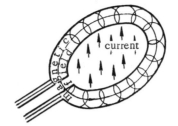

Fig. 6. Magnetic probe.   Fig. 7. Rogowsky loop.

probe, which is a wire loop introduced into the plasma (Fig. 6). A changing magnetic field excites a current in the circular loop and this current is measured on an oscilloscope. Knowing the distribution of the magnetic field and using the laws of electrodynamics one can calculate the distribution of currents in the plasma. In order to measure the current density directly a Rogowsky loop (Fig. 7) is used; the plasma current generates a field that produces a current in the Rogowsky coil, which is wound around the plasma.

The magnetic field can also be measured by a spectroscopic method based on the Zeeman splitting of spectral lines. It is difficult to use this method in a plasma because of the broadening of the spectral lines due to pressure effects and particle collisions, which tend to obscure magnetic field splitting. For this reason spectroscope methods are generally not used for measuring magnetic fields in laboratory experiments. These methods do, however, yield all of our knowledge of the magnetic field in stellar plasmas.

The light emitted in the magnetic field is used to measure weak magnetic fields in the sun. This light is polarized differently in the various components into which the spectral lines are split.

By rotating a polarizer, which accepts light in only one plane of polarization, variations in intensity are observed near the edges of spectral lines in the magnetic field. These variations provide evidence that the lines are split into components with different directions of polarization. However, this splitting is smeared by the broadening of the lines. The range of the variation of the light intensity in rotation of the polarizer is a measure of the magnetic field in the plasma. A remarkable device is based upon this principle. It is called a solar magnetograph and can measure weak varying magnetic fields at the surface of the sun. This method is not suitable for laboratory measurements since the luminous volume is much too small and the changes in the condition of the plasma are much too rapid.

Quasi Neutrality and Charge Separation

A plasma is a mixture of positively and negatively charged particles, ions and electrons, in which the negatively charged electrons are almost completely neutralized by the positively charged ions. We call such a mixture quasi-neutral, i.e., almost neutral. What "almost" means in the present case will now be discussed.

All charge separations give rise to electric fields and in any dense plasma these fields become very large. In accordance with the laws of electrostatics, if we have a volume charge density q over a length x, then there is an electric field which we can express as $E = qx/\varepsilon_0$ in rationalized MKS units. The quantity $\varepsilon_0 = 8.86 \cdot 10^{-12}$ farad/m is called the vacuum permittivity. The field E is measured in volts per meter. Let there be Δn "excess electrons" in 1 m^3 over and above that which neutralizes the ion charge. Then the density of volume charge is

$$q = e\Delta n,$$

where $e = 1.6 \cdot 10^{-19}$ is the electron charge in coulombs. Thus, the electric field arising due to the charge separation is

$$E = 1.8 \cdot 10^{-8} \Delta n x \quad \text{V/m}.$$

We will take as an example a plasma with the same particle density as atmospheric air near the surface of the earth, $2.5 \cdot 10^{-19}$ molecules or $5 \cdot 10^{19}$ atoms/cm^3 ($5 \cdot 10^{25}$ atoms/m^3). Now we convert this air into a plasma in such a way that each atom loses one electron. Thus, our newly converted plasma contains only singly charged ions. The electron density is

$$n = 5 \cdot 10^{25} \text{ electrons/m}^3.$$

The electron density is assumed to change by 1% over a length of 1 cm. Thus, $\Delta n = 5 \cdot 10^{23}$ electrons/m^3, x = 1 cm, and due to this charge separation the electric field which arises is

$$E = 9 \cdot 10^{13} \text{ V/m}.$$

From this example it is clear that in dense plasmas the charge separation must be extremely small.* In the example above, if the electric field is not to exceed some reasonable value, say 90,000 V/m, the charge separation must be smaller than a billionth of a percent! This example illustrates clearly the property of quasi neutrality.

The density of positive and negative particles satisfies the condition of quasi neutrality to a high degree of accuracy. But as seemingly insignificant as the violation of quasi neutrality is, the disturbance arising from it can be very large.

As our example we chose a very dense plasma. However, tenuous plasmas, in which charge separation leads to much smaller electric fields, are more often encountered. Typically, for laboratory experiments the plasma can be assumed to have a density of $\sim 10^{18}$ electrons/m^3 in which 1% charge separation over a length of 1 cm gives rise to an electric field of "only" 1,800,000 V/m. We see that even for such reasonably tenuous plasmas over a length of 1 cm the condition of quasi neutrality is satisfied to a high degree of accuracy. But on a microscopic scale charge separation can become significant. We then also deal with small time differences. In this way quasi neutrality implies electrical neutrality on the average for sufficiently large lengths or time intervals.

A very important role in plasma physics is played by the spatial scale of charge separation. This is that length below which (in order of magnitude) charge separation can become significant. We see that the larger the length over which separation occurs, the stronger the resulting electric field. But for the establishment of an electric field energy is necessary. If the electrons are displaced by a distance x and the electric field is qx/ε_0, then each electron receives an energy $qx^2 e/2\varepsilon_0$. In the absence of external forces this energy must come only at the expense of the thermal energy. The charge separation must spontaneously occur

---
* Indeed the creation of such a strong field requires an immense amount of energy; this matter is discussed below.

over a length x if the electrostatic energy does not exceed the thermal energy:

$$\frac{qx^2 e}{2\varepsilon_0} \leqslant \frac{m\bar{v}^2}{2},$$

where $\bar{v}$ is the average thermal speed of the electrons. Hence,

$$x \leqslant \frac{\bar{v}}{\sqrt{\frac{qe}{\varepsilon_0 m}}}.$$

The first part of this inequality is also the spatial scale of the charge separation. To have complete charge separation all the electrons must be displaced so that $\Delta n = n$, $q = en$. Therefore, the space scale of the charge separation, which we will denote by h (not to be confused with Planck's constant), is

$$h = \frac{\bar{v}}{\sqrt{\frac{ne^2}{\varepsilon_0 m}}}.$$

The time in which particles moving with mean thermal speed cover this distance is the **time scale for charge separation**:

$$t = \frac{h}{\bar{v}}.$$

Quasi neutrality can be violated for times smaller then this value. A disturbance of neutrality leads to a rapid oscillation of the charge density. These oscillations are so characteristic of plasmas that we call them **plasma oscillations** (although there are possible many other kinds of oscillations in a plasma), and the associated frequency is called the **plasma frequency**. Since quasi neutrality is violated during a plasma oscillation, the frequency of the oscillations must be of order $1/t$, where t is the time scale for charge separation. When we study the two-fluid model we will find that the reciprocal of the time scale is just the circular frequency of the plasma oscillations:

$$\omega_0 = \sqrt{\frac{ne^2}{\varepsilon_0 m}}.$$

The ordinary frequency is

$$f_0 = \frac{\omega_0}{2\pi} = 8960 \sqrt{n},$$

where f_0 is the number of oscillations per second, and n is the number of electrons in 1 cm^3. Plasma oscillations are also called electrostatic* oscillations.

We have examined the scale of charge separation for the electrons. The electrons, being the mobile particles, are primarily responsible for the charge separation. However, we can also examine the charge separation due to ion displacement.

A plasma can consist of particles of various species, which we will label by the index k. Each particle species is characterized by a mass M_k and a charge number Z_k so that the particle charge is given by $Z_k e$, where e is the electronic charge. In all plasmas one of the types of particle is always the electron, for which M = m and Z = -1. The remaining particles are ions, which we will denote by an index i. They have a mass M_i and charge $Z_i e$. The condition of quasi neutrality can be written in the form

$$\sum_k Z_k \bar{n}_k = 0$$

or

$$\sum_i Z_i \bar{n}_i = \bar{n}_e.$$

Here n is the density of the corresponding particle, i.e., the number of particles per unit volume; Σ denotes summation over the indicated index, and the bar over a quantity denotes the average over time or space.

Each particle species has its own plasma frequency. The circular plasma frequency ω_0 for particles with charge number Z and mass M is

$$\omega_0^2 = \frac{nZ^2 e^2}{\varepsilon_0 M}.$$

---

* This type of oscillation was first investigated by Irving Langmuir, the father of plasma physics. In his honor they are sometimes called Langmuir oscillations.

The time scale for charge separation for particles of a given species is $1/\omega_0$; the space scale is

$$h = \frac{\bar{v}}{\omega_0}.$$

The denser the plasma, the higher the plasma frequency and the smaller the scale for charge separation both in time and in space. Dense plasmas are almost always electrically neutral. The motion of particles in the plasma takes place in such a way that the ions cannot be separated from the electrons; this motion is termed collective. In a tenuous plasma, on the other hand, the space scale for charge separation can become large compared to the dimensions of the plasma volume. In this case the particles move independently of each other. The condition of quasi neutrality is not satisfied and such a system is really not a plasma. We shall use the following definition of a plasma: a plasma is a system of charged particles with a total charge equal to zero in which the space scale for charge separation is much smaller than the dimensions of the plasma.

Polarization of a Plasma

Any force which acts differently on electrons than it does on ions results in a plasma current. However, a constant stationary current can flow in a plasma in only two cases: if it is closed inside the plasma (a current loop), or if it can flow into external conductors (electrodes) which are in contact with the plasma. If the current does not satisfy one of these conditions it causes charge separation and the establishment of an electric field, i.e., plasma polarization. Thus the internal electric field acting inside a plasma can be very different from the externally applied field. This situation gives rise to many paradoxes that are characteristic of plasmas. In the absence of external conductors which close the currents, or nonstationary oscillatory processes, the electric field must stop the current completely.

The situation is simple if the polarization stops the current along the magnetic field or if there is no magnetic field. We can assume that the current is carried only by the electrons since they are much lighter and, hence, more mobile than the ions. In this case the total force on the electrons is the electric field plus the force due to the electron pressure. In plasmas which are not uniform in density or temperature the lack of equilibrium in the electron pressure in the absence of closed currents and external conductors gives rise to an electric field which is directed from the region of low electron pressure to the region of high electron pressure. These effects are similar to electrical effects in metals.

In all metals there are free electrons with definite electron pressures. If two metals with different electron pressures are joined together an electric polarization field arises at the junction. This is called the contact potential. In metals the electron pressure depends on the electron density and is essentially independent of temperature.* In plasmas, the polarization arises from any inhomogeneity in the electron pressure, which can also be caused by a lack of equilibrium in the electron temperatures. In the absence of a magnetic field the electric polarization field is hardly noticeable. It does not excite currents; rather it stops all possible currents. Polarization does not give rise to any other effects and, therefore, is usually not considered, although it can be very large in stars.

The polarization field becomes more significant in the presence of a magnetic field. In this case the electric field, which is at right angles to the magnetic field, causes the plasma to move. This motion is called a drift, and we will have more to say about drifts later. In a magnetic field all internal fields associated with the polarization become significant. Since all plasma motion perpendicular to the magnetic field excites currents, polarization plays an important role in the dynamics of magnetoplasmas.

In thinking about plasmas, polarization forces us to change our usual ideas of causal relationships between various phenomena. In electrical engineering we are accustomed to thinking that the electric field (or voltage) is the cause and the current is the effect. The magnitude of the current is found by multiplying the external electric field by the conductivity of the material (Ohm's law). In turn, the current is regarded as the cause of the magnetic

---

* Thermoelectric phenomena in metals and plasmas are associated with transport processes.

field (e.g., in an electromagnet). In plasma physics things are the other way around. We must usually assume the magnetic field to be the cause; it determines the velocity and the electric current. The distribution of the electric field is a consequence of this causal chain. Thus, the equation that describes the equilibrium or motion of a plasma is usually transformed such that the electric field is eliminated. This procedure is particularly evident in magneto-hydrodynamics. Ohm's law is not used as an equation to find the current as a function of a given electric field, but rather to find the electric field from a known current (if necessary). The conductivity of the plasma not only determines the magnitude of the current that flows in it, but also the scattering or dispersion of energy in the plasma, i.e., the conversion of ordered motion into thermal motion.

Gas Discharges

When an externally applied electric field ionizes a gas and excites an electrical current in the resulting plasma, we have a gas discharge. In a classical gas discharge the current flows between conducting metal electrodes. In recent times various forms of electrodeless discharges have assumed greater and greater importance. In an electrodeless discharge the current is excited by a time-varying magnetic field. Closed currents flow in electrodeless discharges; these currents are independent of polarization. In this section we will only consider classical electrode discharges.

The electrodes establish an electric field in the plasma. The charge separation caused by this field produces polarization of the plasma. In order to have a stationary current flow through the plasma the space charge that arises in the plasma must be compensated by electrons that come from external sources.

Since the negative electrons are much more mobile than the positive ions, in the presence of the applied field they move to the positive electrode (anode) and the plasma column between the electrodes becomes positively charged. In order for current to flow under these conditions it is necessary that the negative electrode (cathode) inject electrons into the plasma. The injection of electrons by a solid body is called emission. Special means for exciting cathode emission must be employed in order to obtain a discharge at low voltages. We can shine light with a sufficiently short wavelength on the electrode as a means of knocking electrons out of it (photoelectric effect), or we can heat the cathode to a high temperature (thermal emission). Such a discharge, maintained by external means, is called a nonself-sustaining discharge. If the voltage between the electrodes is high enough the cathode can emit electrons without any external agency. This type of discharge is said to be self-sustaining. There are a variety of emission mechanisms. In a dense gas, at very high voltages the cathode is simply heated by the ions that strike it. In this case the emission is thermal, as in the nonself-sustaining discharge with a hot cathode. Such discharges are called arcs (or electric arcs). In a rarefied gas at moderate voltages various forms of cold or glow discharges are possible. Here the cathode emits electrons by a mechanism called field emission. The electric field near the cathode surface extracts electrons directly from the metal. An additional role can be played by secondary electron emission (knocking electrons out of a metal by the collision of ions on the cathode surface). Only if the emission is unlimited can the plasma polarization be completely cancelled by the electron current from the cathode. In a self-sustained discharge the emission is not unlimited so that the plasma column far from the cathode retains a positive charge. It is called the positive column. The application of a voltage primarily affects the region near the cathode, where it facilitates electron emission; this region is called the cathode fall.

Plasma Thermodynamics

Thermodynamics is the science which concerns itself with the properties and behavior of bodies in a state of thermal equilibrium. In a dense medium interparticle collisions lead to the rapid establishment of equilibrium conditions. Conversely, in a tenuous plasma, in which collisions are rare, large deviations from equilibrium can persist for a long time. As we shall see below, in fully ionized plasmas the probability of collisions between particles decreases rapidly with increasing temperature. Thus, we can say that dense, cold, and (in particular) weakly ionized plasmas are in thermal equilibrium as a rule. The properties of such plasmas can be described by thermodynamics. On the other hand, tenuous, fully ionized hot plasmas can exist for relatively long times in a nonequilibrium condition. In these cases a thermodynamic description of the plasma state is not valid.

From the point of view of thermodynamics temperature is a quantity that characterizes the energy distribution of the particles (in the case of a plasma, the electrons and ions). In thermal equilibrium the energy distribution is given by a smooth function; this function was first described by Maxwell. This distribution is called a Maxwellian (Fig. 8). It shows what portion of the total number of particles have energies lying in a given interval. The temperature enters into this function as a parameter, i.e., it determines the shape of the curve. The higher the temperature, the greater fraction of the particles that have high energies. In particular, the temperature determines the average value of the thermal energy per degree of freedom of the particle motion. If the temperature is measured in terms of an absolute scale (the Kelvin scale, °K), then the average energy per degree of freedom is (1/2)kT, where k is the Boltzmann constant (the universal gas constant divided by Avogadro's number).

For simple particles, not having internal degrees of freedom, the entire thermal energy is associated with translational motion; each particle has three degrees of freedom (motion in three mutually perpendicular directions). The average energy of such a particle is (3/2)kT. In fully ionized plasmas the electrons and ions can only perform translational motion (they have no internal degrees of freedom). Therefore, the thermal energy of a fully ionized equilibrium plasma is expressed very simply:

$$E = \frac{3}{2} kT$$

(here E is the energy per particle).

If the energy of interparticle interaction is small compared to the thermal energy, then the plasma is known in thermodynamic

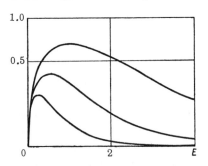

Fig. 8. Maxwellian distribution.

terms as an ideal gas. In this case the plasma energy is almost identical with its thermal energy and the pressure is found from the equation of state for an ideal gas

$$P = (n_e + n_i) kT,$$

where n_e and n_i are the electron and ion densities.

In atomic physics the electron volt (eV) is taken as the unit of energy. It is the energy which an electron acquires in falling through a potential difference of 1 V. Numerically it is very small: 1 eV = $1.6 \cdot 10^{-19}$ joule. The number of electrons in a measureable quantity of matter is very large. If each atom in one gram-molecular weight (mole) of a material has 1 eV of energy, then in thermal units this gram-molecular weight possesses 23,050 calories of energy.

Temperature is also conveniently measured in energy units. We measure the temperature in terms of the quantity kT, which characterizes the energy of thermal motion of the atoms and molecules.* When temperatures are given in energy units thermodynamic formulas are simplified since the Boltzmann constant vanishes. The energy of translational motion per particle is expressed simply as 3/2 T and the specific heat becomes a dimensionless number. The electron volt is taken as the energy unit of temperature. For conversion to degrees we use the relation 1 eV = 11,600°K.

Temperatures necessary for realizing thermonuclear reactions are so high that it is convenient to introduce a unit a thousand times larger, the kilo-electron volt (keV): 1 keV = $1.16 \cdot 10^7$°K.

The plasma pressure is found from the relation

$$p = 1.6 \cdot 10^{-19} (n_e + n_i) T,$$

where T is the temperature (eV), n_e and n_i are the electron and ion densities in particles per m^3, and the pressure p in MKS units is in N/m^2. In order to convert the pressure to practical units (atmospheres, atm, or torr) it is necessary to use a conversion multiplier:

1 atm $\approx 10^5$ N/m^2,
1 torr $\approx 1/760$ atm $\approx 1.3 \cdot 10^2$ N/m^2.

---

* Speaking more precisely, the energy corresponding to two degrees of freedom in the motion of the material particles.

Thus,

$$p = 1.6 \cdot 10^{-24} (n_e + n_i) T \text{ atm} \approx 1.22 \cdot 10^{-21} (n_e + n_i) T \text{ torr}.$$

Normal room temperature is 300°K ≈ 1/40 eV. For this temperature

$$p = 3.1 \cdot 10^{-23} (n_e + n_i) \text{ torr}.$$

Let us now investigate the use of these simple formulas. In very low density plasmas which are not in thermal equilibrium the concept of temperature is not meaningful. Under these conditions, it is a mistake to speak of a plasma temperature. This example can be illustrated from the history of thermonuclear research. A few years ago an apparatus called ZETA (Zero Energy Thermonuclear Apparatus) was constructed in England to investigate heating and confinement of plasmas. It was first believed that high temperatures would be reached in this device. The temperatures were measured by measuring the broadening of spectral lines emitted by heavy ions, these being present in the plasma as an impurity. However, measurements on the lines of various ions yielded "temperatures" which were higher for the more highly charged ions. Evidently, the line broadening resulted from ion motion, caused by acceleration in the electromagnetic fields. The energy acquired as a result of this motion was proportional to the ion charge. In plasmas which are not in thermal equilibrium, the various particles move with different energies and the concept of temperature for such a plasma is not defined.

At higher densities a condition known as **partial thermal equilibrium** is possible. Thus, in a plasma in a magnetic field, the velocities parallel and perpendicular to the field can be distributed according to different laws, although the distribution for each of these directions can be approximately an equilibrium Maxwellian. Under these conditions we can say that the plasma has two temperatures: one, corresponding to motion along the field, is called the **longitudinal temperature** and the other, corresponding to motion at right angles to the field, is called the **transverse temperature**. Such phenomena, in which the properties of a medium are different in different directions, are called **anisotropic**. We commonly denote the direction parallel to the magnetic field with a subscript $\parallel$, and the direction perpendicular to the field by a subscript $\perp$. In this way the longitudinal temperature is indicated by $T_\parallel$, and the transverse temperature by $T_\perp$. Each of these temperatures can be used to define a pressure in its direction: $p_\parallel \neq p_\perp$. Therefore, tenuous plasmas located in a magnetic field exhibit an **anisotropic pressure**. As the density increases energy exchange between the various degrees of freedom is facilitated and the anisotropy of the temperature and the pressure is equalized.

The most difficult process of all is energy exchange between electrons and ions; this is because of the large difference in mass. Electrons that strike ions bounce off them like tennis balls bouncing off a massive object, and almost no energy is transferred. Therefore, the electron temperature in a plasma T_e can differ from the ion temperature T_i. A plasma with different electron and ion temperatures is an example of a system in partial thermodynamic equilibrium. It can be regarded as a mixture of two gases; the electron gas and the ion gas. Each of these exist in thermal equilibrium (the velocity distributions for both the electrons and the ions are Maxwellian); however, the two gases are not in equilibrium with each other.

At very high densities all plasmas rapidly approach a state of **total thermal equilibrium** in which the electron and ion temperatures are equal. The time required to establish this equilibrium is called the **relaxation time**; we will have more to say about this below (thermodynamics only investigates states and not the process by which they are established). At still higher densities the thermodynamic properties of the plasma can change; the plasma no longer behaves as an ideal gas.

These departures from the ideal gas law are associated with two new phenomena which occur at high densities, electrostatic interactions and degeneracy.

At high densities the plasma energy is not only determined by the thermal energy of the particles, but also by the potential energy of their interaction. In a plasma each charged particle is surrounded by an "atmosphere" of surplus particles of opposite charge which shield the electrostatic field of the particle. The distance over which this screening is effective is of the order of the space scale for charge separation.

In a plasma, consisting of particles of different charge, the shielding length h (called the Debye or polarization length) is found from the charge separation scales by adding their inverse squares:

$$\frac{1}{h^2} = \sum_k \frac{1}{h_k^2},$$

where h_k is the separation scale:

$$h_k = \frac{v}{\omega_0} = \frac{v_k}{\sqrt{\frac{Z_k^2 n_k e^2}{\varepsilon_0 M_k}}}.$$

In a thermal plasma it is necessary to take the average thermal velocity of the particle for v:

$$\bar{v}_k \approx \sqrt{\frac{T}{M_k}} \; ; \quad h_k^2 = \frac{\bar{v^2}}{\omega_0^2} = \frac{\varepsilon_0 T}{Z_k^2 e^2 n_k}.$$

Here n_k is the density of particles with charge number Z_k; T is the temperature in energy units. Therefore, in an equilibrium plasma

$$\bar{h} = \frac{\sqrt{\varepsilon_0 T}}{e} \cdot \frac{1}{\sqrt{\sum_k n_k Z_k^2}}.$$

The summation is carried out over all particles, including both ions and electrons, for which we write Z = -1. We can introduce the quantity

$$\bar{Z} = \frac{\sum_k n_k Z_k^2}{n},$$

where $n = n_e + n_i$ is the total density of particles.

From the condition of quasi neutrality Z is nearly equal to the charge number of the ions. From this expression the shielding length is found to be

$$\bar{h} = \sqrt{\frac{\varepsilon_0 T}{n \, e^2 Z}}.$$

The shielding length is larger, the higher the temperature and the lower the density of the plasma. If we express the temperature in convenient practical units, i.e., electron volts and the density in the number of particles in 1 m³, the shielding length in meters is

$$\bar{h} \approx \frac{1500}{\sqrt{\bar{Z}}} \sqrt{\frac{T}{n}}.$$

In a dense plasma the shielding length is extremely small even for high temperatures. Thus, the plasma energy is reduced by the mutual attraction of all the particles and their surrounding "atmospheres" with excess opposite charge. The electrostatic energy per particle is

$$\Delta E = -\frac{e^3}{8\pi}\left(\frac{\bar{Z}}{\varepsilon_0}\right)^{3/2} \sqrt{\frac{n}{T}}.$$

This is called the **electrostatic energy** and is negative just like any attraction energy. The electrostatic energy can be thought of as the energy that arises if all the oppositely charged particles in the plasma attract each other at a distance equal to the shielding length. In practical units the energy can be written

$$\Delta E \approx 1.0 \cdot 10^{-13} (\bar{Z})^{3/2} \sqrt{\frac{n}{T}} \text{ eV},$$

and the ratio of the electrostatic energy to the thermal energy is

$$\frac{\Delta E}{T} \approx 1.0 \cdot 10^{-13} \sqrt{n} \left(\frac{\bar{Z}}{T}\right)^{3/2}.$$

The plasma can be described thermodynamically as an ideal gas as long as this quantity is small;

$$\frac{\Delta E}{T} \ll 1,$$

i.e., for sufficiently low densities or high temperatures. The limiting density arises when the corrections for the electrostatic energy exceed 1%. This limiting density is

$$n - 10^{22}\left(\frac{T}{\bar{Z}}\right)^3.$$

The limiting density is proportional to the cube of the temperature; thus, the corresponding pressure goes as the fourth power of the temperature. Therefore, if $\bar{Z} \approx 1$ (a hydrogen plasma) at a temperature of 1 eV the electrostatic interaction can be neglected if the plasma density is less than 10^{22} particles/m^3 and the pressure, a hundredth of an atmosphere. At a temperature of 10 eV this is the case up to a density of about 10^{25} particles/m^3 and a pressure of the order of 30 atm. Cold dense plasmas rapidly depart from the ideal gas law; by "cold" we mean temperatures less than 10,000°K.

A reduction in the plasma energy is accompanied by a corresponding pressure reduction, which can be calculated using the equations of thermodynamics. As long as the electrostatic energy is small in comparison with the thermal energy, the reduction in the pressure is given by

$$\Delta p = \frac{1}{3} \Delta E \cdot n.$$

All of our equations for electrostatic interaction are derived from the theory of Debye. For this reason the shielding length h is called the Debye length and a sphere of radius h is known as a Debye sphere. We can say that the electrostatic forces are important in the volume of the Debye spheres around each particle. Debye's theory is statistical and can only be used when the number of particles in the Debye sphere is large. Using the equations presented above we can easily convince ourselves that the number of particles in a Debye sphere is inversely proportional to the ratio of the electrostatic energy to the thermal energy:

$$\frac{4}{3} \pi \bar{h}^3 n = \frac{1}{6} \cdot \frac{T}{\Delta E}.$$

Therefore, all the formulas given for corrections to the energy and pressure due to the electrostatic interaction are useful for determining the densities for which the thermodynamics of ideal gases can be applied. Going to this limit we enter a region in which the plasma behaves like a compressed gas or liquid heated above the critical temperature. No simple laws of thermodynamic plasma properties can describe this region. In general, it is not practical to investigate this region since the calculation of the mutual collective interaction of many particles is very difficult. However, it is easy to find asymptotic laws of plasma thermodynamics for high densities. These laws relate to the region where the plasma energy and pressure are not determined by the electrostatic interaction, but by a different physical phenomenon, degeneracy.

A degenerate electron gas appears as a direct consequence of one of the basic principles of quantum physics, the Pauli principle. This principle states that not more than one electron can occupy each quantum state. In order to apply this principle to a gas it is convenient to employ a six-dimensional phase space, the volume being the product of the usual coordinate volume $\Delta V = \Delta x \cdot \Delta y \cdot \Delta z$ and the volume in momentum space $\Delta p_x \cdot \Delta p_y \cdot \Delta p_z$. Here p is the particle momentum, i.e., the product of the mass and the velocity. Each quantum state occupies an elementary cell in phase space with volume h^3, where h is Planck's constant. Each cell cannot contain more than two electrons (and these must spin in opposite directions).

If the density is small then the volumes in phase space are only sparsely occupied by electrons and the Pauli principle has little effect on the properties of the plasma. But if the plasma is highly compressed a situation can arise in which all cells are occupied. With further compression the electrons are "squeezed out" in momentum space into cells with higher energies. At very strong compressions the energy and pressure of the electron gas increases independently of the temperature. The thermodynamic relationship for the limiting case of complete degeneracy is obtained very simply. In this condition all cells up to the maximum momentum p_{max} are occupied. If the plasma filling ordinary space has a volume V, then the volume in phase space is $^4/_3 \pi p_{max}^3 V$ ($^4/_3 \pi p_{max}^3$ is the volume of a sphere in momentum space). The number of elementary cells in this volume is

$$\frac{4}{3} \pi \frac{p_{max}^3}{h_3} V.$$

Each cell can contain two electrons with opposite spins. Therefore, the total number of electrons with momenta less than

p_{max} that can be accommodated in phase space is

$$N = \frac{8}{3}\pi \frac{p_{max}^3}{h^3} V.$$

The density, i.e., the number of electrons per unit volume is

$$n = \frac{N}{V} = \frac{8}{3}\pi \frac{p_{max}^3}{h^3}.$$

Thus, if the electron density is n, the maximum momentum is

$$p_{max} = h\sqrt[3]{\frac{3}{8\pi}n},$$

and the maximum kinetic energy is

$$E_{max} = \frac{p_{max}^2}{2m} = \frac{h^2}{2m}\left(\frac{3}{8\pi}n\right)^{2/3}.$$

A simple calculation shows that if the electrons occupy a sphere in momentum space with uniform density, the average energy per electron is

$$E = \frac{3}{10}\cdot\frac{h^2}{m}\left(\frac{3}{8\pi}n\right)^{2/3}$$

and, in accordance with the laws of thermodynamics, the pressure is

$$p = \frac{2}{3}\overline{E}n = \frac{1}{5}\cdot\frac{h^2}{m}\left(\frac{3}{8\pi}\right)^{2/3} n^{5/3}.$$

This is also the limiting equation of plasma thermodynamics for very high densities. The energy E_{max} is called the degeneracy energy or the Fermi energy. If it is large in comparison with the thermal and electrostatic energies, the energy and pressure of the plasma will be determined by the energy and pressure of the degenerate electron gas. It is interesting that in this case the pressure is proportional to the density raised to the 5/3 power just as in the adiabatic compression of an ideal gas. However, the energy and pressure in a completely degenerate plasma do not depend on the temperature.

Elementary Processes

The transition of a gas into the plasma state involves various processes of interparticle interaction. These processes involve collisions of particles between themselves or interactions of particles with radiation. First we should mention the process called ionization, i.e., the stripping of electrons from atoms or molecules. Without ionization it would be impossible to obtain a plasma. Ionization occurs in two ways: in a dense plasma, by electron collisions, and in a highly rarefied plasma, by the action of radiation (light, ultraviolet, or x-rays). In principle, ionization can also be accomplished by atom − atom or atom − ion collisions, but considerably higher energies are necessary.

The inverse process to ionization is called recombination (combining of ions and electrons to form neutral atoms or molecules). Ionization is a threshold process: the energy of the colliding particle or radiation photon must be higher than some threshold value, which depends on the stability of the atom. This threshold value is called the ionization energy. Recombination requires the inverse condition: the atom being formed must rid itself of excess energy or else it disintegrates rapidly. Where excess energies are to be disposed of, it is necessary to differentiate between two completely different recombination processes: one of these is radiative recombination, the other is three-body recombination. In the latter case two electrons must collide simultaneously with an ion; one of the electrons attaches itself to the ion and the other carries away the excess energy. In tenuous plasmas radiative recombination is the dominant process. However, the probability of the emission of a light photon by collision is small; in the majority of cases the colliding particles are scattered and only exchange energies. This type of collision is called an elastic collision. In a dense plasma, recombination takes place mainly through the three-body process.

Ionization can occur in both atoms and molecules. The result of ionizing a molecule is a molecular ion; this particle subsequently disassociates into atomic ions and neutral particles.

Of the various elementary processes in plasma physics, the process called charge exchange is one of the most important.

In this process an ion, colliding with an atom, acquires an electron from the atom. The ion becomes an atom and the atom an ion (see Fig. 40). Charge exchange provides a basis for energy loss by charged particles because of the transfer of the energy from the ions to neutral atoms. In dense partially ionized plasmas another, less important, elementary process is possible. An atom can capture an extra electron and be turned into a negative ion. Oxygen atoms have an especially large affinity for electrons. The formation of negative oxygen ions is important in the physics of the upper atmosphere, where electrons are captured by oxygen atoms. This process tends to reduce the electrical conductivity of the ionospheric plasma.

We will not dwell on chemical processes such as the dissociation of molecules into atoms. We mention only that chemical processes occur in plasmas. For example, hydrogen-containing molecules can attract hydrogen ions (protons). The molecular ion that results can exist only in a charged condition; if the charge is removed it breaks up quickly. In this way hydrogen molecules can produce H_3^+ ions, methane molecules can produce CH_5^+, etc. However, such processes require a high concentration of neutral particles and, thus, are not very relevant to plasma physics.

Plasmas and Radiation

The northern lights, lightning, and the advertising lights on a busy city street — are all evidence that plasmas are luminous, i.e., they radiate. In addition to radiating visible light, plasmas can radiate invisible ultraviolet rays. Hot plasmas can also radiate in the x-ray region of the spectrum. All of these emissions are similar, differing only in frequency (or wavelength). However, there are a number of different radiation processes: discrete, recombination, and bremsstrahlung. Each of these forms of radiation can occur in the various spectral regions.

Discrete radiation consists of individual spectral lines. Each line arises as a result of electron transitions from one energy level to another within an atom. This radiation is also known as bound — bound radiation. All other forms of radiation have a continuous spectrum. Recombination radiation is emitted in the capture of a free electron by an ion having charge number Z; an ion with smaller charge, or a neutral atom is formed. Bremsstrahlung (a German word for braking radiation) is emitted in weaker interactions of a free electron with an ion in which the electron is not captured, but only decelerated. Recombination radiation is described as free − bound, bremsstrahlung as free − free.

At low temperatures in weakly ionized plasmas most of the radiation is discrete. Undissociated molecules, which exist in the plasma because of the low temperatures, radiate in broad molecular bands, instead of lines. At higher temperatures the role played by continuous spectra, recombination, and (for still higher temperatures) bremsstrahlung is more important. If total ionization is achieved and all the electrons have been liberated from all of the atoms (which can most easily be done in a hydrogen plasma in which each atom has only one electron to lose), the discrete lines vanish entirely. On the other hand, if there are heavy atoms in the plasma then even at high temperatures there are multiply charged ions with electron shells. These ions radiate considerable energy in the form of line radiation in the far ultraviolet. For this reason the contact of hot plasma with material walls has catastrophic consequences. With the evaporation of wall material, heavy atoms enter the plasma and radiate a sizable amount of energy. This energy causes further evaporation of the walls; as a result the plasma is not only cooled, but is also saturated with impurities. Recombination and bremsstrahlung radiation increase sharply with increasing ion charge. The presence of multiply charged ions in the plasma increases the energy loss into these forms of radiation. Consequently, to heat plasmas to high temperature (as required for thermonuclear reactions) the plasma must contain only light atoms (with small nuclear charge), preferably hydrogen or its isotopes, deuterium and tritium.

The basic theory of radiation can be stated as follows: the more intensely a body can absorb radiation, the more intensely it can emit the same radiation. Therefore, in a thick layer, a plasma is opaque for all radiation which it can radiate. Each radiation process has a corresponding inverse absorption process. For discrete radiation the inverse process is discrete absorption of lines. The inverse process for recombination is the absorption of light which leads to ionization of the atoms. This process is called the photoelectric effect or photon absorption. Finally, bremsstrahlung radiation corresponds to the inverse process of bremsstrahlung absorption.

A beam of light, propagating through an absorbing medium, is attenuated in a geometrical progression. The mathematical expression for the attenuation of the beam intensity J is given by

$$\ln \frac{J_0}{J} = kx,$$

where x is the thickness of the layer through which the beam propagates. The coefficient k is called the **absorption coefficient**, and its reciprocal is called the **absorption half length**.

The absorption of radiation in the plasma is not irreversible. This radiation imparts energy to the electrons which reradiate it, but in another direction. Such absorption with subsequent reemission is equivalent to scattering and leads to radiative **diffusion**. Since radiation transfers energy in this way, radiative diffusion is also called **radiative thermal conductivity**. Radiative diffusion is slower, the larger the absorption coefficient. However, the same process, absorption with subsequent emission, can occur in **true scattering of light** by electrons (this is called **Compton or Thomson scattering**). Optical absorption depends very strongly on the frequency of the light, but electron scattering is essentially frequency independent.

The over-all coefficient, characterizing both the absorption averaged over all frequencies and the scattering is called the **opacity** of the plasma. Radiative diffusion (radiative thermal conductivity) depends only on the opacity of the plasma. By multiplying the opacity by the thickness of the plasma layer we can obtain a nondimensional quantity called the **optical thickness**. A plasma layer with large optical thickness is opaque to radiation. Radiation emerges from this layer only by the slow process of multiple reemission and scattering. This radiation is said to be **trapped** and is in thermal equilibrium with the material. In this

case a plasma satisfies the condition of detailed balance*
and maintains thermal equilibrium of ionization. Conversely, a
plasma layer with small optical thickness is transparent to radiation. The radiation emerges freely from such a layer and the
concepts of radiative diffusion and radiative thermal conductivity
lose their meaning. Thermal equilibrium is not realized either between the radiation and the material, nor between ions, electrons,
and neutral atoms.

The more dense a plasma, the greater its opacity. In a dense
plasma a thin layer is "optically thick"; in a tenuous plasma a thick
layer is "optically thin." Therefore, a dense plasma is almost always in thermal equilibrium and a rarefied plasma is very often
not in equilibrium.

A hot plasma in a magnetic field radiates magnetobremsstrahlung or synchrotron radiation. In a magnetic
field the thermal motion of the plasma particles is a combination
of unrestrained motion along the magnetic field lines and cyclotron gyration around them. The gyration of the electrons around the magnetic field lines leads to magnetic radiation. This
form of radiation only occurs in very hot plasmas which contain
relativistic electrons, i.e., electrons whose velocities
are not small compared to the speed of light. Relativistic electrons are not produced in a cyclotron, but in another kind of accelerator known as a synchrotron. Therefore, the magneto-bremsstrahlung radiation from relativistic electrons is called synchrotron radiation. This form of plasma radiation is very important in astrophysics. Radiotelescopes detect radio waves arriving
here on earth from outer space. One of the chief sources of this
radiation appears to be synchrotron radiation from gaseous nebulae. A strong source of synchrotron radiation is the Crab Nebula,
the remains of a stellar explosion which was recorded by Chinese
astronomers in the year 1054. When thermonuclear research is
able to produce a fully ionized hot plasma which does not contain
multiply charged ions (thereby eliminating all other radiation
losses) the only known obstacle to increasing the temperature
will be synchrotron radiation.

---

* The condition of detailed balance implies that the direct and inverse processes go through the same channel.

Equilibrium and the Stationary Ionization States

If an incompletely ionized plasma is maintained under constant external conditions the simultaneous action of the ionization and recombination processes lead to the establishment of conditions in which the rate of ionization is equal to the rate of recombination. Thereafter, the densities of ions and electrons remain constant. A condition of this kind is said to be **stationary**.

In many important cases the stationary condition coincides with the condition of **thermodynamic equilibrium**. Equilibrium is established if the direct and inverse processes take place by identical mechanisms, i.e., the principle of **detailed balance** is obeyed. If the ionization is due to electron collisions the inverse process of recombination must be due to three-body recombination. If, on the other hand, the ionization is caused by photons, recombination must occur with the emission of identical photons.

The principle of detailed balance is always satisfied in **closed systems**, in which there is no exchange with the surrounding medium. In a system of this kind the radiation is in equilibrium with matter; as a result the stationary ionization condition coincides with the equilibrium condition.

In **open systems** the stationary condition coincides with equilibrium only if the density of the plasma is sufficiently high. In dense plasmas both ionization and recombination proceed basically in the same way: ionization by electron impact, and recombination by the three-body process. In rarefied plasmas this cannot be true. At low densities three-body collisions are not very probable. Recombination occurs with the emission of radiation. But while the radiation escapes freely from the system the principal ionization process is the electron collision. Thus, in these open systems the principle of detailed balance is not obeyed and the stationary ionization condition does not coincide with thermodynamic equilibrium. If the plasma is surrounded by an enclosure which is opaque to radiation the radiation is stored in the system until complete thermodynamic equilibrium is established. However, this is not practical for hot tenuous plasmas. The **solar corona** is a plasma which exists under the conditions described above.

Quantitatively ionization is characterized by the **degree of ionization** α, i.e., the fraction of ionized particles relative to the initial number of particles. In a fully ionized plasma the degree of ionization approaches unity; in a weakly ionized plasma it is a small fraction. In thermodynamic equilibrium the degree of ionization depends only on the temperature and density of the plasma. In a rarefied plasma in which the stationary condition of ionization cannot coincide with equilibrium the degree of ionization depends on how much radiation emerges from the system.

In all plasmas except pure hydrogen, ionization takes place gradually. First, the most weakly bound electrons are detached, then the next, etc.

The condition of thermodynamic equilibrium for each degree of ionization is expressed by the Saha equation

$$\frac{n_l}{n_{l-1}} = \frac{2G_l}{G_{l-1}} \cdot \frac{1}{\tilde{n}_e} e^{-I/T} \approx \frac{1}{\tilde{n}_e} e^{-I/T}.$$

Here n_i is the density of the ions with the greatest charge and n_{i-1} is the density of ions with less charge (or neutral atoms); $\tilde{n}_e$ is the average number of electrons in an elementary cell of phase space; I is the ionization energy of the ion* with less charge; T is the temperature in energy units. Usually I and T are expressed in electron volts; then the ionization energy I is numerically equal to the ionization potential. The coefficients G_i and G_{i-1} are the so-called statistical weights of the corresponding ions (or atom and ion). Taking account of excited levels we can write the statistical weight in the form

$$G = g_0 + g_1 e^{-E_1/T} + \dots .$$

Higher terms of this series are usually unimportant. Here g_0, g_1,... are the statistical weight constants of the zeroth, first, etc. energy states of the atom or ion; E is the excitation energy (or potential). The statistical weight of a quantum state with S for the

---

*If an atom has several electrons, by removing one of these we obtain an ion from which further electrons can be removed. The energy required for this is the ionization energy of the ion.

spin and L for the orbital angular momentum is

$$g = (2S+1)(2L+1).$$

The coefficient $2G_i/G_{i-1}$ is usually of order unity and is usually taken to be unity.

In the first stage of ionization the ion with smallest charge is a neutral atom and the Saha equation assumes the form

$$\frac{n_i}{n_a} \approx \frac{1}{\tilde{n}_e} e^{-I/T},$$

where n_a is the density of atoms; n_i is the density of singly charged ions; I is the ionization potential of the atoms.

The average number of electrons in an elementary cell of phase space can be written

$$\tilde{n}_e \sim n_e \lambda^3,$$

where n_e is the electron density; λ is the quantum mechanical wavelength of the electron (also known as the de Broglie wavelength)

$$\lambda = \frac{h}{m\bar{v}} = \frac{h}{\sqrt{2\pi mT}}.$$

Here, $\bar{v}$ is the average thermal speed of the electrons in one direction

$$\bar{v} \sim \sqrt{\frac{T}{2\pi m}};$$

and T is the temperature in energy units.

Expressing $\tilde{n}_e$ in terms of n_e we can write the Saha equation in the form

$$\frac{n_i \cdot n_e}{n_{i-1}} \approx \frac{1}{\lambda^3} e^{-I/T}.$$

Thus, we see that the Saha equation is a special form of the relation that describes chemical equilibrium. It is obtained from the usual law of mass action if we regard the ions and electrons as chemical materials, and ionization as the inverse chemical reaction.

Using the Saha equation it is easy to estimate the approximate temperature required for each stage of ionization to occur. Let $T_{1/2}$ denote the temperature for 50% ionization, i.e., the temperature for which

$$n_{i-1} = n_i.$$

Using the Saha equation, we find

$$T_{1/2} = -\frac{I}{\ln \dfrac{1}{\tilde{n}_e}}.$$

According to the Pauli principle the number of electrons in an elementary cell is never greater than one.* Thus, the quantity in the denominator of the logarithm is positive. If n is approximately one, then a degeneracy arises and the Saha equation becomes inapplicable. In the usual nondegenerate plasma $\tilde{n}_e$ is much smaller than unity and $T_{1/2}$ is several times smaller than I.

The form of the Saha equation presented above is convenient for multiply charged ions present in a plasma as an impurity. One of the methods of plasma diagnostics consists of observing the spectral lines emitted by these ions. The appearance of lines from multiply charged ions can provide an estimate of the temperature if the principle of detailed balance obtains. The electron density can be considered as given, since it is determined by the ionization of the basic components of the plasma, and not by the small impurities.

In order to calculate the electron density in a plasma we write the equation in a slightly different form. If only singly charged ions are present in the plasma then the condition of quasi neutrality tells us that the ion density is equal to the electron density

$$n_i = n_e = n.$$

---

* The number of electrons in an elementary cell can be either one or zero. The average number is therefore less than unity.

In this case the Saha equation becomes

$$n \approx \frac{\sqrt{n_a}}{\lambda^{3/2}} e^{-I/2T}.$$

The temperature for 50% ionization remains the same as in the preceding case. In this form the Saha equation applies to hydrogen plasmas and to other plasmas in the first stage of ionization, as long as the degree of ionization is small. We assume that the density of atoms has not changed and our equation becomes

$$\alpha \approx \frac{1}{\sqrt{n_a \lambda^3}} e^{-I/2T}.$$

If the stationary condition of ionization does not coincide with equilibrium, we cannot use thermodynamics to find the degree of ionization but must equalize the ionization and recombination rates.

Plasmas as Conducting Fluids

A plasma is a fluid medium capable of conducting electrical current. Therefore, a simple model for a plasma is the model of a conducting fluid. In this model we do not examine the motion of individual particles; instead, the plasma is assumed to be a continuous medium similar to those that are treated in hydrodynamics. When this simplifying assumption is used there are no differences between liquids and gases. In the absence of ionization the difference between these states of matter lies in the compressibility, gases being much more compressible. However, compressibility is also important in flow phenomena if the flow velocity is near the sound speed. Therefore, ordinary hydrodynamics is the study of the motion of liquids and gases at velocities which are small compared with the speed of sound.

Plasmas, i.e., conducting gases, differ from conducting liquids not only in compressibility but also in other degrees of freedom associated with charge separation. For this reason the hydrodynamic model has a narrow range of applicability: the flow velocity must be small and the oscillation frequencies must be small. However, slow processes in a plasma can be described by the con-

PLASMAS AS CONDUCTING FLUIDS 45

ducting fluid model. In this model detailed differences between plasmas and liquid metals, such as mercury or liquid sodium, are not taken into account. As a further simplification we usually assume ideal conductivity, i.e., completely negligible electrical resistance in the plasma or, in other words, infinite conductivity. This limiting case accentuates the features that distinguish conducting fluids from ordinary fluids. All of the relations assume an especially simple form. It is convenient to begin our investigation with the case of an ideal conductor; later we shall introduce corrections for the finite conductivity.

The motion of a conducting fluid is determined by the effect a magnetic field has on it. Therefore the study of conducting fluids is called magnetohydrodynamics.

The approach used in magnetohydrodynamics is based on combining the solutions of the equations of hydrodynamics and electrodynamics. We will not give the complicated system of equations, but state the final results. These can be summarized in the following three basic concepts of magnetohydrodynamics:

a) freezing of the magnetic field;
b) magnetic pressure;
c) diffusion of the magnetic field.

The first two relate to the limiting case of an ideal conductor, and the third establishes the limits of this approximation.

Freezing of the magnetic field is a direct consequence of the law of electromagnetic induction for ideal conductors. In accordance with the law of induction, if a conductor moves across magnetic field lines an electromotive force is induced in it. However, in an ideal conductor, i.e., one with infinite conductivity, the smallest electromotive force can produce an infinitely large current, which is clearly impossible. Consequently, the motion of the ideal conductor must be such that magnetic field lines are not intersected by the motion. If an ideal conductor is forced to move across the magnetic field then it must drag the magnetic field lines along with it.

If an electric field acts on a conductor in such a way as to induce a current, and if the conductor is ideal, it must move across the magnetic field with a velocity such that the induced electromotive force cancels the electric field. This motion is called a drift, and the velocity is called a drift velocity. In the presence of drift motion the magnetic field lines move together with the conductor. Thus, freezing of the field is compatible with the laws of drift motion. We will discuss the drift motion of the various particles in the plasma in more detail in terms of the independent particle model. We will then see how the drift velocity is produced.

To this point we have only been concerned with wire conductors. However, in speaking of fluids, we can think of a "fluid wire" consisting of certain definite particles of matter. The "freezing" relation says that if the conductivity is ideal, then when the fluid wire moves across a magnetic field it cannot intersect a field line. This means that the field must move with the fluid into which it is frozen. In other words we can say that the magnetic field lines "stick" to the material particles. If the freezing relation obtains the fluid motion at each point adjusts itself to the local drift velocity.

The second basic relation of magnetohydrodynamics is closely associated with the first. According to the equations of electrodynamics, if the strength of the magnetic field varies in a direction perpendicular to the magnetic field lines, then an electric current must flow in the direction perpendicular to both directions. However, a conductor with a current flowing at right angles to a magnetic field experiences a force proportional to the field, which is called the pondermotive force. If the current density is known, then the value of this force can be calculated easily. It is proportional to the product of the magnetic field B and the current density $j_\perp$ that flows perpendicular to the field. In the MKS system of units the absolute value of this force is given by

$$|F| = j_\perp B.$$

The vertical bars indicate the magnitude of the force; the direction is not specified.

If we express the current in terms of the magnetic field then the current can be ignored and the same force can be treated as a magnetic pressure force rather than a pondermotive force.

The laws of electrodynamics tell us that if the magnetic field is parallel to the z-axis and the current density j flows along the x-axis, then the magnetic field changes in the y-direction according to the relation

$$\frac{dB}{dy} = \mu_0 j.$$

The current in this equation is the sum of the ordinary conduction current and the so-called displacement current, which was introduced into electrodynamics by Maxwell. However, if we do not treat fast oscillations, the displacement current can be neglected and we can assume that j is the conduction current. Now let us substitute our expression for the current in terms of the derivative of the magnetic field into the expression for the pondermotive force

$$|F| = \frac{1}{\mu_0} B \frac{dB}{dy} = \frac{d}{dy}\left(\frac{B^2}{2\mu_0}\right).$$

If we are interested in the direction of the force as well as its magnitude then these equations must be rewritten in vector form:

$$F = j \times B,$$

$$\nabla \times B = \mu_0 j;$$

thus,

$$F = \frac{(\nabla \times B) \times B}{\mu_0} = \frac{(B \cdot \nabla) B}{\mu_0} - \nabla\left(\frac{B^2}{2\mu_0}\right).$$

If the magnetic field varies only at right angles to its own direction, then

$$F = -\nabla \frac{B^2}{2\mu_0}.$$

The operator ∇ denotes the gradient, i.e., the derivative with respect to direction.

Calculations show that the pondermotive force is always in the direction of decreasing magnetic field. The pressure force is expressed in similar fashion if we assume that the role of a pressure is played by the quantity

$$p_M = \frac{B^2}{2\mu_0}.$$

It is convenient to treat this quantity as a **magnetic pressure**, and to treat the pondermotive force as the force associated with the magnetic pressure.

The ratio of the gas-kinetic pressure to the magnetic pressure is an important characteristic of a plasma. This dimensionless number is denoted by the Greek letter β:

$$\beta = \frac{p}{p_M} = \frac{2\mu_0 nT}{B^2}.$$

The concepts pondermotive force and magnetic pressure are two ways of expressing the same physical effect. Both have the same validity and lead to identical results.

The same idea can also be arrived at from other points of view. In electrodynamics it is known that the magnetic field establishes a pressure equal to the magnetic energy density, i.e., proportional to the square of the magnetic flux density:

$$p_M = \frac{B^2}{2\mu_0}.$$

If B is expressed in teslas, then the pressure will be expressed in N/m^3, i.e., approximately in terms of a hundred thousandth of an atmosphere. It should be realized that this pressure does not act on nonconducting bodies, therefore, it is not taken into account in ordinary hydrodynamics. As a matter of fact a magnetic field can penetrate nonconducting bodies freely. The pressure can only act on an impenetrable barrier. From the field-freezing concept discussed above it follows that the magnetic field cannot penetrate an ideal conductor in the direction perpendicular to the field lines. This means that the entire force of the magnetic pressure must act on the ideal conductor in this direction.

In treating the motion of an ideally conducting fluid or gas (plasma) in addition to considering the usual kinetic pressure, we must also take the magnetic pressure into account. In doing this, one should note two complicating circumstances. First, the magnetic pressure acts only in the direction perpendicular to the magnetic field lines, i.e., it is **anisotropic**. Second, to a large degree the magnetic pressure is effective only in the limiting case of an ideal conductor. If the conductivity is finite the field-freezing picture is violated and the magnetic field gradually "leaks" through the surface of the body. The body feels the magnetic pressure only for a time which is small compared with the leakage time, or, as it is usually called, the diffusion time for the magnetic field. For longer times the magnetic pressure is equalized and ceases to act just as air pressure cannot act on a porous screen. The rate at which the magnetic field "leakage" occurs is determined by the third law which we call the law of magnetic field diffusion. It is obtained by solving the equations of magnetohydrodynamics together with the usual Ohm's law, which says that the current density j is proportional to the electric field E:

$$j = \sigma E.$$

The coefficient of proportionality σ is called the conductivity. Hence, the equation for leakage of the magnetic field should resemble the ordinary diffusion equation. The magnetic field diffusion coefficient D_M is inversely proportional to the conductivity σ; as $\sigma \to \infty$ for an ideal conductor, $D_M \to 0$. In the MKS system of units the magnetic field diffusion coefficient is given by

$$D_M = \frac{1}{\sigma \mu_0}.$$

The diffusion coefficient, as in the usual case, has the dimensions m²/sec. Magnetic field diffusion is subject to the usual laws governing random processes. The penetration depth in a time t is of order

$$t \approx \sqrt{D_M t} \approx \sqrt{\frac{t}{\sigma \mu_0}}.$$

In a small time t the magnetic field (and therefore the current) can only penetrate through a thin surface layer of the conductor (thickness l), which we call the skin depth. From another point of view, the field and current penetrate a distance L in a time

$$t \approx \frac{L^2}{D_M} \approx \mu_0 \sigma L^2,$$

which we call the skin or diffusion time. An alternating current penetrates into a conductor only during a half period of the cycle. Therefore, the high-frequency current only flows in a thin skin layer near the surface of the conductor. The depth of this skin layer can be obtained if we replace t by a quantity of the order of the reciprocal of the frequency (usually we take $t = 2/\omega$, where ω is the angular frequency).

Field Diffusion and Plasma Diffusion

The diffusion time is the time during which the magnetic field penetrates into a stationary conductor. If the plasma is confined by the pressure of the magnetic field, diffusion of the magnetic field into the plasma violates the pressure balance and the plasma begins to move. To be more specific we should not speak of magnetic field diffusion into the plasma, but rather of plasma diffusion into the magnetic field. If the ratio of the kinetic pressure to the magnetic pressure $\beta = p/p_M$ is small, then the difference in the magnetic pressure inside and outside the plasma is small. The time needed for the field to diffuse into the plasma is equal to the skin or diffusion time given above and is determined by this small difference. However, total equilibrium requires a much longer time, since the plasma must be uniformly distributed over the entire region which it originally did not occupy. This plasma diffusion time is not determined by small pressure differences, but by the total magnetic pressure; it is $1/\beta$ times larger than the skin time. Correspondingly we can assume that the plasma diffusion coefficient is many times smaller than the magnetic field diffusion coefficient.

The law of magnetic field diffusion establishes the region in which the assumption of perfect conductivity can be applied. The dimensions of the system must be large compared with the thickness of the skin layer and the time must be small in comparison with the diffusion time in the skin layer. These scales are associated with the conductivity, but a knowledge of the conductivity does not solve the problem. A conductivity can be "large" or "small" depending on the characteristic length and time scales.

Applications of the Conducting-Fluid Model

Length scales in space are so large that they are considerably greater than the skin depth for processes which are reasonably fast. Therefore, in space situations all plasmas can be assumed to have infinite conductivity and the concepts of field freezing and magnetic pressure generally apply. Using these laws we can solve a number of problems in astrophysics. For example, a plasma stream known as the solar wind is ejected from the sun and impinges on the upper atmosphere of the earth with significant effects. This stream has no magnetic field embedded in it. According to the field-freezing concept, an external magnetic field should be unable to penetrate such a plasma. There is a random interplanetary magnetic field in the solar system. The plasma stream arriving from the sun displaces this external magnetic field. It might be said that the "plasma broom" sweeps the interplanetary magnetic field from the vicinity of the sun. Magnetic belts are formed around the sun in which the magnetic field is weaker than in neighboring regions. The magnetic belts facilitate the passage to the earth of fast charged particles which are ejected from the sun (corpuscular stream). In other words, when it encounters the magnetic field of the earth, the plasma stream flows around it the way a liquid flows around a solid body. A magnetic cavity is formed around the earth inside of which a magnetic field is confined but which the plasma stream cannot enter. Many charged particles collect on the surface of this magnetic cavity. These particles are observed by satellites and rockets as the outer radiation belt. The analogy of the streaming of the plasma around the cavity and the flow of a liquid around a solid body is more than qualitative; the two phenomena can be described by similar equations. However, in the plasma case the magnetic pressure must be taken into account in addition to the kinetic pressure.

Elementary concepts of plasma confinement by magnetic fields can also be understood on the basis of the conducting-fluid model. Thermonuclear reactions require temperatures so high that no walls can be exposed to them. The plasma can reach an equilibrium without walls if its kinetic pressure is balanced by the pressure of an external magnetic field

$$nT = \frac{B^2}{2\mu_0},$$

where n is the number of plasma particles per unit volume and the temperature is in energy units. If we use the usual energy unit of atomic physics, the electron volt (1 eV = $1.6 \cdot 10^{-19}$ joules = 11,600°K), then the confinement condition is

$$B = 6.34 \cdot 10^{-13} \sqrt{nT},$$

where T is the temperature in these energy units. Reasonable parameters for controlled thermonuclear reactions are n = 10^{22} particles/m^3 and T = 10^4 eV. Under these conditions a magnetic field of about 6 tesla is required for confinement. This is a technically feasible field strength.

The confining magnetic field can be established either by external currents in metal conductors wound around the plasma or by internal currents in the plasma. Correspondingly, we distinguish between confinement by external and intrinsic magnetic fields.

The simplest external field is a purely longitudinal field established by a coil (solenoid) which surrounds the plasma (see Figs. 9 and 15). The simplest intrinsic field is a circular field produced by longitudinal currents flowing in a plasma (see Fig. 25). It follows from the magnetic-pressure concept that the longitudinal current flowing in the plasma forces the plasma to be compressed into a thin cylindrical filament by the action of its own circular magnetic field.

Such a plasma filament is usually known in the English and American literature as a "pinch."* The concept of plasma com-

---

* In the recent literature the word "pinch" has been used broadly to describe plasma configurations. Depending on the direction of the electric current we have various possibilities; for example, the "linear z-pinch," in which the current flows along the axis of the cylinder (z-axis), and the magnetic compression field is azimuthal. There is also the θ-pinch (theta-pinch), in which an azimuthal current flows around the plasma cylinder in the θ-direction, and the compression magnetic field is longitudinal. In the θ-pinch the current can only be excited by induction so that the actual compression is produced by a magnetic field which is established by currents flowing in external conductors. On the other hand, the z-pinch is established by passing a current between electrodes which are introduced into the plasma. When one speaks of a simple pinch, one actually means the classical z-pinch.

THE CONDUCTING-FLUID MODEL

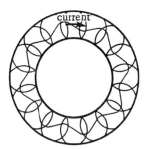

Fig. 9. A toroidal magnetic bottle.

pression by intrinsic fields acting on currents flowing in the plasma is called the pinch effect.

Therefore, even with this crude model we can make an estimate of the conditions needed for plasma containment. However, to refine these conditions and to investigate specific methods of confinement we must use a more sophisticated model of a plasma. We shall first examine the conditions which provide plasma confinement until the magnetic field begins to leak into the plasma, i.e., times less than the skin diffusion time. To estimate the skin diffusion time it is necessary to know the conductivity of the plasma and this requires a somewhat more detailed description.

Toroidal Plasma Traps

A magnetic field keeps the plasma from drifting to the side walls, but has no effect on motion toward the ends of a tube. A simple way of producing a magnetic trap is to make an endless tube. We simply bend the tube into a circle, as shown in Fig. 9. A body similar to a doughnut or bagel has a special geometric name, the torus. A magnetic trap of this form is referred to as toroidal. An external coil or solenoid generates a magnetic field along the tube. This field keeps the plasma from drifting in the transverse direction into the walls. Although the plasma moves freely along the field this motion does not result in plasma loss. The magnetic field lines are closed into circles and do not intersect the walls of the bottle. However, plasma confinement in a toroidal trap is complicated by the fact that the magnetic field cannot be homogeneous in such a trap.

A homogeneous field is one in which the lines are straight and everywhere equally spaced. In a torus the field lines are curved of necessity, and are compressed toward the inner wall of the torus. The spacing of the magnetic field lines determines the intensity of the magnetic field and, consequently, the magnetic pressure. Since the field lines are compressed toward the inner wall of the torus, we know that the magnetic pressure will be greater there than near the outer wall. This difference or g r a d i e n t in the magnetic pressure forces the plasma toward the outer wall. Thus, even our simple model of a plasma allows us to understand the basic problem of confining a plasma in a toroidal trap. This problem can be examined in more detail by using an independent particle model, as we shall see later. The drift of plasma to the walls is associated with the d r i f t m o t i o n of the particles in the inhomogeneous field. The centrifugal force that acts on particles that move along the curved field lines is also important. This effect is in the same direction as the field gradient effect.

To confine a plasma in a torus it is necessary to use a more complicated magnetic field. All the methods that have been for this purpose proposed amount to the formation of a h e l i c a l m a g n e t i c f i e l d. The simplest method can be used where a strong electric current flows through the plasma and along a magnetic field. The current is excited in the plasma inductively; the plasma acts as the shorted secondary of a transformer whose primary is an annular conductor that carries an alternating current. The current flowing in the plasma sets up a circular magnetic field around itself which, in combination with the external longitudinal field, forms the helical field. If the self-field is stronger than the external field the plasma pinch is compressed and is detached from the walls; this is called a t o r o i d a l p i n c h. However, it is also possible to have a trap in which the plasma is isolated from the walls by a strong external longitudinal field and the intrinsic field only plays a subsidiary role. Traps with currents that flow along the magnetic field are subject to various hydromagnetic and kinetic instabilities; these will be discussed in a later section. Thus, it is useful to devise a trap in which the helical field is established by currents that flow in external conductors; this type of trap is called a s t e l l a r a t o r.* In its original form the stellarator was a torus twisted into a figure eight. In later models the builders learned to construct a helical field without distorting the torus through the use

---

* The name "stellarator" comes from the word "stellar" (starlike). This name is used because it is hoped to achieve temperatures of the same order as those found in the regions of stars in which thermonuclear reactions occur.

of auxiliary windings with current flow in opposite directions in alternating windings.

Plasma can be heated in a toroidal magnetic trap. The magnetic field not only prevents the drift of the plasma to the walls, but also reduces the energy loss to the walls by reducing the transverse thermal conductivity of the plasma, as we shall see below. If there are no multiply charged ions the energy loss due to radiation is small. Then in principle it should be possible to heat plasmas to temperatures such that thermonuclear reactions can occur. The practical solution to the problem of heating plasmas in a magnetic trap involves a number of difficulties, however. The simplest scheme is to heat a plasma by a current flow along the magnetic field. However, the current can "drive" the plasma, making it unstable. We are therefore forced to seek other heating methods in which the current flows perpendicular to the magnetic field. To this end, use can be made of resonances between the interacting plasma and high-frequency electromagnetic waves, in particular the cyclotron and magnetoacoustic resonances.

Electromagnetic Pumping and Plasma Acceleration

In addition to confining plasmas, the magnetic pressure can be used to accelerate plasmas to high velocities. There are a number of methods for projecting individual plasmoids (or plasma blobs) or continuous plasma jets from a nozzle by using the magnetic pressure. As in the confinement problem, a magnetic field can be created by currents, flowing either in external conductors or in the plasma.

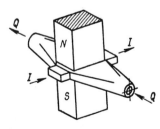

Fig. 10. Conduction electromagnetic pump.

The essence of the plasma acceleration process can be understood on the basis of the conducting-fluid model. In fact, the same principle is applied in electromagnetic pumping, which is used to transport molten metals. The simplest device of this type is the conduction electromagnetic pump (Fig. 10). In this pump a channel with a conducting fluid is placed between the poles N and S of a magnet. A current I is passed through the fluid at right angles both to the magnetic field and the channel axis. An interaction takes place between the mutually perpendicular current and magnetic field and the pondermotive force drives the fluid stream Q in the direction shown in the figure. The constant current is generated by an external electric field and is applied to the fluid through electrodes. This is a conduction current and is the reason why the pump is called a conduction pump. There are also electromagnetic pumps in which the current is not introduced through electrodes, but is excited (induced) by an alternating magnetic field. Such pumps are called induction pumps. As an example we can consider a device in which a conducting medium is carried along by a traveling magnetic field. This scheme is shown in Fig. 37. If the scheme is used to transport a molten metal the device is called an inductive electromagnetic pump. In plasma acceleration applications this is called an asynchronous plasma motor. In plasma acceleration the conduction pump is called a crossed-field accelerator.

In plasma accelerating devices electrical energy is transformed into mechanical energy. The process is analogous to that which occurs in an electric motor. Acceleration in crossed fields uses the same principle as a constant-current motor; acceleration by traveling fields invokes the principle of the asynchronous motor. However, it is known that an electric motor can be converted into a generator by a simple reversal. The same can also be done with plasma motors if they are used to decelerate plasmas. In this case the kinetic energy of the plasma stream is transformed into electrical energy. We then obtain a plasma generator, or, as it is usually called, a magnetohydrodynamic generator.

The schemes which we have discussed thus far employ magnetic fields established by currents in the special windings. However, there are also schemes in which the magnetic field is generated by currents that flow in the plasma itself. Such schemes are variations of what is known as the rail accelerator. The name derives from a simple scheme in which the plasma is accelerated between two parallel straight conductors (or rails) which carry a current that is closed through the plasma. In a symmetric system the self-magnetic field only compresses the current pinch. However, in the rail system the field is applied to one side by

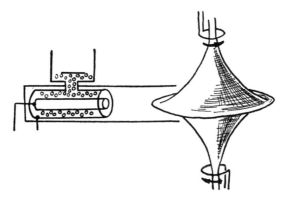

Fig. 11. A plasma gun, injecting plasma into a cusp trap.

metal conductors and flows freely into the other, carrying the plasma along with it.

Injection of a plasma jet into a magnetic trap can be carried out with a **plasma gun** based on the same principle. It is convenient to use a **coaxial** configuration in which an annular gap between two coaxial cylinders serves as a channel for the plasma. A coaxial plasma gun is shown in Fig. 11. It consists of two cylinders with a common axis. The plasma is fed into the annular gap between the cylinders by a fast-acting valve (from above). A radial electric field applied between the cylinders produces a radial current in the plasma which, interacting with its own magnetic field, expels the plasma from the gun. The injection of plasma into a bottle with opposed fields by a gun was suggested to prevent the interchange, or flute, instability, which we shall discuss in a later section.

Magnetohydrodynamic Flow

Plasma flow in a magnetic field is similar to the flow of a fluid or gas in that it can be either **laminar**, i.e., in smooth layers, or **turbulent**, i.e., vortex flow. In most cases a magnetic field directed along the flow inhibits the formation of turbulence, that is to say, it stabilizes the laminar motion. The stabilizing action of the magnetic field is strongest in a highly conducting fluid, in which the field-freezing effect applies. In this case turbulent motion leads to tangling of the field lines as a result of which the kinetic energy of the motion is transformed into magnetic energy. In a weakly conducting plasma the magnetic field also inhibits the growth of turbulence, but the stabilization mechanism is altogether different. It is due to the fact that the motion of a conducting medium in a magnetic field gives rise to an electric current which, for finite conductivity, leads to an energy dissipation, i.e., the kinetic energy of the motion is converted into thermal energy (Joule heat). A similar dissipation process occurs in nonconducting media because of viscosity. For this reason the quantity $1/\sigma\mu_0$ (the magnetic diffusion coefficient) is sometimes called the magnetic viscosity.

The nature of the flow of a nonconducting fluid or gas is determined by the dimensionless Reynolds number

$$Re = \frac{vL}{\nu},$$

where v is the flow velocity, L is a linear dimension, and ν is the kinematic viscosity. At small Reynolds numbers the flow is laminar; at large numbers it becomes turbulent. In magnetohydrodynamics the character of the motion depends on more than one dimensionless number. If we replace the kinematic viscosity by the magnetic viscosity $1/\sigma\mu_0$ in the Reynolds number we obtain the magnetic Reynolds number:

$$Re_M = \sigma\mu_0 vL.$$

If the magnetic Reynolds number is large the flow medium can be assumed to be infinitely conducting; if Re_M is small the conductivity is low.

In the limiting case of large magnetic Reynolds numbers the characteristics of the motion are determined by the ratio of the magnetic pressure $B^2/2\mu_0$ to the pressure head $\rho v^2/2$:

$$A = \frac{B^2}{\mu_0 \rho v^2}.$$

For large values of the parameter A (greater than 0.1) the

magnetic field completely stabilizes the laminar flow of a highly conducive plasma regardless of the Reynolds number.

In the limiting case of small magnetic Reynolds numbers the character of the flow is determined by the dimensionless Stuart number

$$S = ARe_M = \frac{B_0^2 \sigma L}{\rho v}.$$

If this number exceeds 0.1 laminar flow is stable for all Reynolds numbers.

All of the criteria given above apply to flows along the magnetic field. The nature of the flow across a magnetic field in a channel of width L is determined by the value of the dimensionless Hartmann number

$$M = \sqrt{SRe} = BL\sqrt{\frac{\sigma}{\rho v}}$$

For small values of the Hartmann number the flow is the same as with no magnetic field. At large values of M the viscosity is only effective in a thin layer (of order L/M) at the walls of the channel. In the rest of the cross section the velocity is constant regardless of the distance from the wall. In this case the resistance to the motion does not depend on viscosity, being determined entirely by the electromagnetic forces.

The Two-Fluid Model

In the conducting-fluid approximation the current density is found from Ohm's law, which is not very accurate for a plasma. A more detailed description of the electric currents in a plasma is given by the two-fluid model, the two fluids being the electrons and the ions. The motion of each of these species is described by the usual hydrodynamic equations which as we have seen are known as the hydrodynamic approximation, in contrast with the more accurate physical kinetics description.

The two-fluid model allows us to obtain additional results,

including some complicated ones. However, because of the inaccuracy of the hydrodynamic approximation not all of these are correct. We shall limit ourselves to certain simple and reliable cases.

In the two-fluid model it is assumed that all the electrons at a given point move with the same velocity v_e, and all the ions with the same velocity v_i (we are speaking of the ordered velocities; thermal motion requires the method of physical kinetics to be treated properly). It is convenient to define the **mass velocity**

$$v = \frac{Mv_i + mv_e}{M + m} \approx v_i$$

and the **relative velocity**

$$v_i - v_e \approx v - v_e,$$

which is associated with the **current density**

$$j = Zen_i v_i - en_e v_e,$$

where Ze is the ion charge, and e is the electron charge.

The difference between the total ion and electron charge must be very small because otherwise the space charge produces an enormous electric field. Therefore, the condition of **electrical neutrality**

$$Zn_i \approx n_e \equiv n,$$

holds and the current density is proportional to the relative velocity

$$j \approx ne(v_i - v_e).$$

In terms of the mass velocity the two-fluid model produces nothing new. Therefore, we investigate the results in terms of the relative velocity, i.e., the current.

A simple result is obtained if we assume that all velocities are constant in time (the stationary case) and that there is no magnetic field. In this case the two-fluid model leads to a result iden-

THE TWO-FLUID MODEL

tical with Ohm's law, which means that the conductivity σ can be identified. In the absence of a magnetic field the ions necessarily move much more slowly than the electrons so that the current is carried almost entirely by electrons, which are accelerated in the electric field and decelerated as a result of collisions with ions. In the stationary case the acceleration vanishes, i.e., the momentum obtained by the electrons from the electric field must be equal to the momentum lost in collisions with the ions. The momentum obtained by the electrons from the E field per unit time is eE. The momentum imparted to the ions is conveniently written as

$$\frac{m}{\tau}(v_e - v_i),$$

where τ is the average time between collisions.* Equating these quantities, we obtain

$$v_i - v_e = \frac{e\tau}{m} E,$$

from which the current density is

$$j = \frac{ne^2}{m} \tau E.$$

This result agrees with Ohm's law, with the conductivity being identified as

$$\sigma = \frac{ne^2 \tau}{m}.$$

The quantity τ is determined by investigating collisions between the plasma particles.

It is important to note that Ohm's law applies to plasmas only when there is no acceleration or magnetic field. We now turn to nonstationary processes, i.e., time-dependent plasma processes in which acceleration must be taken into account.

In the absence of an external constant magnetic field, there is only one possible plasma oscillation. This oscillation is purely

---

* Strictly speaking, the deceleration of the electrons is due not only to collisions, but also to the continuous interaction with ions at large distances. We will later examine random processes and shall see the method by which these continuous interactions can be described approximately, which is known as Coulomb collisions.

electrostatic and is associated with charge separation. It is called the **plasma oscillation**.

A simple descriptive model of plasma oscillations can be visualized. Consider a plasma layer of area S and thickness x. Now assume that all nSx electrons are displaced to one of the bounding surfaces of this layer. As a result a parallel-plate capacitor is formed with charge

$$Q = neSx.$$

The capacitance of this layer is $\varepsilon_0 S/x$. Therefore, the potential difference between the "plates" is

$$\Delta\varphi = \frac{xQ}{\varepsilon_0 S} = \frac{nex^2}{\varepsilon_0}.$$

The electric field in the capacitor is

$$E = \frac{\Delta\varphi}{x} = \frac{nex}{\varepsilon_0}.$$

The field arising as a result of the charge separation acts on the electrons with a force

$$F = -eE = -\frac{ne^2 x}{\varepsilon_0}.$$

The minus sign indicates that this is a restoring force. As in a pendulum or a spring, the restoring force F is proportional to the displacement from equilibrium, x.

In the case of a spring the restoring force F = -kx gives rise to oscillations with an angular frequency

$$\omega = \sqrt{\frac{k}{m}},$$

where m is the mass of the weight which is hung from the spring.

In the plasma case the quantity ne^2/ε_0 plays the role of the spring constant k and the role of the weight is played by the electron with mass m. Consequently, if neutrality is violated oscillations occur with an angular frequency given by

THE TWO-FLUID MODEL

$$\omega_0 = \sqrt{\tfrac{4\pi n e^2}{m}}.$$

The ordinary frequency (number of oscillations per second) is

$$f_0 = \tfrac{\omega_0}{2\pi} = 8.960\sqrt{n},$$

where n is the number of electrons per cubic meter. This frequency, as we noted earlier, is called the **plasma** or **Langmuir frequency**.

The plasma frequency is a basic quantitative characteristic of the plasma. It is proportional to the square root of the electron density. The coefficient of proportionality only contains universal constants. For a rough estimate it is useful to remember the following approximate relation:

$$f_0 = 8.960\sqrt{n} \approx 10\sqrt{n}.$$

The accuracy provided by this approximation is more than adequate for present-day experimental methods of determining the density.

It is interesting to estimate the plasma frequency for various phenomena in nature and technology. Radio waves received on the earth from the solar corona are emitted by plasma with a density n of about 10^{14} particles/m³. The corresponding plasma frequency is roughly 100 MHz (megahertz), corresponding to wavelengths in the meter range (ultrashort waves). Still lower values of the density and plasma frequency are found in the ionosphere; the ionosphere is responsible for the propagation of short waves over large distances. Laboratory plasmas have considerably larger densities. In most gaseous discharges the plasma density is of the order of 10^{18} particles/m³, and the corresponding plasma frequency is 10^{10} Hz, i.e., centimeter wavelengths. Controlled thermonuclear reactions will require densities of about 10^{22} particles/m³; the corresponding plasma frequency, 10^{12} Hz, is in the submillimeter range. Stellar interiors are composed of plasmas which are many orders of magnitude more dense; these plasmas are characterized by plasma frequencies in the infrared region of the spectrum.

If we assume that the electrons are stationary, then the ions can oscillate relative to the electrons at the ion plasma frequency. This frequency is given by the same expression as that for the electrons, except that the electron mass m and charge e are replaced by the ion mass M and charge Ze. In a hydrogen plasma the ion plasma frequency is about 40 times smaller than the electron frequency. In all other cases the difference is larger still. Ion plasma oscillations can be observed when the ions are much colder than the electrons.

Many quantities which characterize plasmas can be expressed in terms of the plasma frequency. In particular, the plasma conductivity is related to the plasma frequency by the expression

$$\sigma = \omega_0^2 \, \varepsilon_0 \tau .$$

In the presence of a magnetic field the application of the two-fluid model means that we must examine the motion both of electrons and ions since it is impossible to say beforehand which particle will move faster. Taking these things into account we find that many new degrees of freedom appear in the plasma.

Plasma Conductivity in a Magnetic Field

If the plasma is located in a magnetic field then the two-fluid model becomes quite complicated. Current flowing along the magnetic field is subject to Ohm's law with the usual value of the conductivity. In plasmas located in a magnetic field this ordinary conductivity is called the longitudinal conductivity. The transverse conductivity is smaller than the conductivity in the direction of the magnetic field. If the transverse current is carried only by electrons which move under the influence of an externally applied electric field the plasma possesses an anisotropic conductivity. The rules for anisotropic conductivities are as follows: first, the conductivity at right angles to the field is reduced in inverse proportion to the square of the magnetic field; second, the current does not only flow parallel to the applied electric field, but also at right angles to it (Hall effect). A plasma in which the electrons perform several cyclotron gyrations between collisions is called a magnetoplasma. A plasma magnetized in this way must have a density that is not too high and must be located in a strong magnetic field. If the anisotropy of the conductivity is significant, in a magnetoplasma the transverse conductivity will be much smaller than the longitudinal conductivity and the current perpendicular to the electric field (the Hall current) will be considerably larger than the current parallel to the electric field. However, in a highly ionized plasma the anisotropy of the conductivity is attenuated by the following three effects. First, the current can be carried at right angles to the field by the ions as well as the electrons. Second, the current transverse to the magnetic field can be produced by forces of a nonelectrical nature, e.g., a pressure differential, in addition to the electric field. Finally, charge separation leads to plasma polarization, i.e., the appearance of an internal electric field in the plasma. Thus, the current will not only be determined by the externally applied electric field, but also by polarization fields.

The more neutral particles there are in the plasma, the stronger the anisotropy in the conductivity. In fully ionized plasmas the diffusion coefficient for the magnetic field and the decay time (skin diffusion time) are always determined by the normal conductivity (i.e., by the conductivity without a magnetic field). In partially ionized plasmas in which there are many neutral particles the skin diffusion time is reduced by the reduction of the transverse conductivity in the magnetic field.*

In addition to the anisotropy in the particle motion anisotropic collisions have an effect on the conductivity in a magnetic field; these are characteristic of highly ionized plasmas. As we shall see later, the faster the particles move, the less often they collide with other particles (for free motion along the magnetic field). In moving across the field, in which case the particles really do not move rapidly, they gyrate around the magnetic field lines until they experience a collision. Because of the anisotropy in the collisions, the conductivity perpendicular to a strong magnetic field in a fully ionized plasma is approximately one half the longitudinal conductivity. Collisions of electrons with neutral par-

---

* If the plasma pressure is small in comparison with the magnetic pressure the plasma diffusion coefficient is equal to the magnetic field diffusion coefficient multiplied by the ratio of these pressures β. The corresponding plasma diffusion time will be $1/\beta$ times larger than the skin diffusion time. In this case the diffusion of plasma (but not of magnetic field) is always determined by the normal conductivity.

ticles depend only slightly on the velocity so that there is very little anisotropy in collisions in a weakly ionized plasma. However, the magnetic field has a much stronger effect on transversely moving particles, which carry the current.

Plasma as an Ensemble of Independent Particles

The direct opposite of the model based on the continuous medium (the conducting-fluid model) is the independent particle model. In this model, in studying the motion of the charged plasma particles, we neglect their mutual interaction completely. Obviously, the independent particle model is more appropriate for the description of tenuous plasmas and the conducting-fluid model is more appropriate for dense plasmas. However, the usefulness of the independent particle model is not restricted to very tenuous plasmas since this model helps us to understand several general plasma properties.

A charged particle with charge Ze and mass M, moving with a velocity v in a magnetic field, is acted upon by a pondermotive or Lorentz force, perpendicular both to the magnetic field and to the velocity of the particle. Under the influence of this force the particle must gyrate around the magnetic field lines with an angular frequency

$$\omega_c = \frac{ZeB}{M}.$$

This is the same frequency with which particles in a cyclotron accelerator gyrate. Therefore, ω_c is called the cyclotron or gyromagnetic frequency. There are two characteristic cyclotron frequencies, the electron gyrofrequency $\omega_{ce} = eB/m$ and the ion gyrofrequency $\omega_{ci} = ZeB/M$. For brevity we will simply refer to these as ω_e and ω_i.

The gyration of a particle in a plane perpendicular to the magnetic field traces out a circle. These orbits are called cyclotron orbits and the radius r_c is called the cyclotron radius. If the magnetic field is directed out of the page, then the particles with positive charge gyrate in the clockwise sense and particles with negative charge gyrate in the counterclockwise sense.

The frequency of gyration along the cyclotron orbit depends only on the strength of the magnetic field and the charge and mass of the particle. All particles of a given type in a given field gyrate with the same frequency. The velocity in this gyration orbit can have any value, depending on the velocity with which the particles enter the magnetic field.

If a plasma is in thermal equilibrium, then the velocities of the gyrations are distributed according to Maxwell's law. In a tenuous plasma, thermal equilibrium is established slowly and the distribution of velocities can be arbitrary, depending on the initial conditions.

The radius of the cyclotron orbit r_c depends on the velocity of gyration

$$\frac{2\pi r_c}{T} = v,$$

where T is the gyration period:

$$T = \frac{1}{\nu} = \frac{2\pi}{\omega_c}.$$

Therefore,

$$r_c = \frac{v}{\omega_c}$$

or after substituting for the quantity ω_c

$$r_i = \frac{M v_i}{Z e B}$$

for the ions and

$$r_e = \frac{m v_e}{e H}$$

for the electrons.

Therefore, for a given velocity the cyclotron radius of the ion is M/Zm times (i.e., a thousand times) larger than the cyclo-

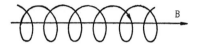

Fig. 12. The helical path of particles in a magnetic field.

tron radius of the electron. If the plasma is in thermal equilibrium the ion and electron energies (not the velocities) are the same:

$$\frac{Mv_i^2}{2} = \frac{mv_e^2}{2},$$

whence

$$\frac{v_e}{v_i} = \sqrt{\frac{M}{m}}.$$

In this case the cyclotron radius of the ions is only $\sqrt{(M/m)}/Z$ times larger than the cyclotron radius of the electrons.

The magnetic field does not affect the motion of particles along the field. Thus, in a homogeneous field and in the absence of other forces the motion of the particles is a combination of the cyclotron gyration and uniform rectilinear motion with a given velocity along the field lines.* The trajectory of the particles under the effect of one pondermotive force is a helix (Fig. 12) whose projection on a plane perpendicular to the magnetic field is the cyclotron circle.

Drift Motion

The motion of particles in the presence of other forces, or in an inhomogeneous field, is rather complicated. On the one hand the motion along the magnetic field lines is accelerated. On the other hand (and this is a remarkable peculiarity of charged particle motion in a magnetic field), a force acting at right angles to the field gives rise to particle motion in a direction perpendicular both to the forces and the magnetic field. This motion is known as a drift.† Drift motion differs from free motion in that the particles do not experience c o n s t a n t a c c e l e r a t i o n under the influence

---
* A homogeneous field is a field in which the strength is constant in magnitude and in direction. Thus, the field lines are straight.

† Sometimes we speak of "drifts perpendicular to the field," in which case the inertial motion along the field can also be called a "drift."

DRIFT MOTION

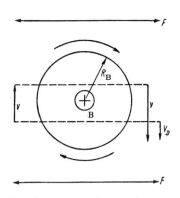

Fig. 13. Diagram showing the cause of drift motion.

of constant forces; rather, they move with a constant velocity.

We have already encountered drift motion in the conducting-fluid model; it appears as a common drift motion of individual particles.

The origin of drift motion can be clarified by the following example (Fig. 13). The force $F_\perp$, acting in the plane of the cyclotron circle imparts an acceleration in one direction for a half period. During the second half period the acceleration is opposite to the direction of gyration. As a result the component of the gyration velocity perpendicular to the force will be larger when the particle moves in one direction (in the figure this is the lower portion), than in the other direction (the upper part of the figure). The net result is that all the cyclotron circles are shifted in the plane of the circle in the direction perpendicular to the force $F_\perp$; in the figure this corresponds to motion downward. The magnitude of the displacement can be estimated easily. Particles of mass M will acquire an acceleration $F_\perp/M$ under the influence of the force $F_\perp$. In the course of one period this acceleration establishes a difference in the "upper" and "lower" velocities:

$$\Delta v \approx \frac{F_\perp}{M\omega_c}.$$

The displacement of the cyclotron circle during a period is of order

$$l \approx \frac{\Delta v}{\omega_c} \approx \frac{F_\perp}{M\omega_c^2}.$$

In unit time the circle is displaced by a distance of order

$$|v_D| \approx l\omega_c \approx \frac{F_\perp}{M\omega_c}.$$

The displacement of the circle per unit time is simply the

drift velocity. We see that it is perpendicular both to the direction of $F_\perp$ and the magnetic field B.

Using the expression for the cyclotron frequency ω_c, we obtain a final expression for the drift velocity:

$$|v_D| = \frac{F_\perp}{ZeB}.$$

The vertical brackets indicate that this expression only gives the magnitude of the velocity, not the direction.

In order to know in which direction the particles drift it is necessary to express this result in vector form.

$$v_D = \frac{F \times B}{ZeB^2}.$$

For electrons $Z = -1$. Thus, if the magnetic field B is out of the page and the force F points upward, positive particles drift to the right and negative particles to the left. The **drift current**, carried both by these and other particles, flows to the right. Although our derivation is approximate it turns out that the results are accurate: in our derivation we neglect a dimensionless factor of order unity which does not enter into the result; the sign for the vector product, which we have not considered, is positive.

It is convenient (as an aid to memory) not to use the drift velocity, but the **drift current** carried by particles of a given species

$$|j_D| = nZev_D = n\frac{F_\perp}{B},$$

or in vector form

$$j_D = n\frac{F \times B}{B^2}.$$

In the case of electrons $Z = -1$ and the direction of the current is opposite to the direction of the drift.

Electric Drift

As our first example we consider the case in which the force F is due to an electric field:

$$F_\perp = ZeE_\perp.$$

Therefore,

$$|v_E| = \frac{E_\perp}{B}.$$

The quantity $E_\perp$ is the component of the electric field perpendicular to the magnetic field. Since the electrons have a negative charge $Z = -1$. The general expression for the drift velocity contains the charge number Z in the denominator. Therefore, if the force $F_\perp$ acts in a similar way on electrons and ions, the effect is a drift of electrons in one direction and ions in the other. We note, however, that the electric field has an opposite effect on electrons and ions; the force is proportional to the charge number Z. Therefore, the charge cancels out of the drift velocity when $ZeE_\perp$ is inserted for $F_\perp$. The drift caused by an electric field is called the **electric drift**. The velocity of the electric drift is the same for electrons and ions, both in magnitude and direction. In vector form it is expressed as

$$v_E = \frac{E \times B}{B^2}.$$

In Fig. 14 we see that if the magnetic field is directed out of the page and the electric field points up, then the particles drift to the right.

The electric drift leads to a drift of the entire plasma, i.e., to a mass motion. It does not excite a current. This is the same drift motion we spoke of earlier in considering the conducting-fluid model. On the other hand, a force such as the force of gravity or a centrifugal force, which acts on all particles in the same way regard-

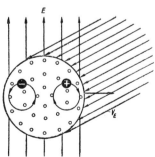

Fig. 14. Electric drift.

less of their charge in the absence of a magnetic field, causes the electrons and ions to drift in opposite directions. In these cases a drift current is excited along with the drift motion.

We see that drift motion possesses paradoxical properties: the forces of electrical and nonelectrical nature act as though they had exchanged roles. The electrical forces cause only a mass motion and the nonelectrical forces excite currents.

In order to distinguish the drift motion from the cyclotron gyration we must have a strong magnetic field; then, in one cyclotron orbit the distortion due to inhomogeneities or other forces is small. The period and radius of the cyclotron gyration will then be small compared with the time t and length L that characterize the change in all other quantities. Motion that satisfies these conditions is called adiabatic motion.*

The quantitative relations that must be satisfied if the motion is to be adiabatic are

$$\omega_c \gg \frac{1}{t};$$
$$r_c \ll L.$$

If F is some quantity which affects the motion, then

$$\frac{1}{t} - \frac{d \ln F}{dt};$$
$$\frac{1}{L} = \frac{d \ln F}{dx}.$$

The expression for the drift velocity can be obtained from the exact equation of motion for particles in a magnetic field:

$$M \frac{d\mathbf{v}}{dt} = Ze(\mathbf{v} \times \mathbf{B}).$$

Here the force F also includes the force due to the electric field ZeE. We resolve the equation into its x and y components. The

---

*The term adiabatic is not used here in the same sense as in thermodynamics, but in a broader sense (as in mechanics). A process is called adiabatic if it takes place with no change (or with slow changes) in the external conditions.

magnetic field is directed along the z-axis. Then

$$\frac{dv_x}{dt} = \frac{ZeB}{M} v_y + \frac{F_x}{M} = \omega_c v_y + \frac{F_x}{M},$$

$$\frac{dv_y}{dt} = -\frac{ZeB}{M} v_x + \frac{F_y}{M} = -\omega_c v_x + \frac{F_y}{M}.$$

We now write this system of equations in complex form:

$$\frac{d}{dt}(v_x + iv_y) = -i\omega_c(v_x + iv_y) + \frac{(F_x + iF_y)}{M}.$$

The solution of this inhomogeneous equation can be written as the sum of the solutions of the homogeneous equation

$$v_x + iv_y = Ce^{-i(\omega_c t + \varphi_0)},$$

which represents the cyclotron gyration, and a particular solution for the inhomogeneous equation v. If the force **F** does not depend on time, then the particular solution is easily obtained by equating the left-hand side to zero; thus

$$v_y = -\frac{F_x}{M\omega_c},$$

$$v_x = \frac{F_y}{M\omega_c},$$

or

$$\mathbf{v} = \mathbf{v}_D = \frac{\mathbf{F} \times \mathbf{B}}{ZeB^2}$$

in agreement with the results which we obtained above by an intuitive method. For a constant force this result is exact. If the force F varies slowly with time this result will be approximately correct; the accuracy improves, the more nearly the motion is adiabatic.

If the motion is adiabatic the particle motion can be analyzed as a combination of three independent motions: free motion along the magnetic field lines, cyclotron gyration around them, and drift motion perpendicular to the magnetic field. If the adiabatic condition is not satisfied the last two motions become confused and a complex pattern arises which is difficult to analyze.

Conservation of Magnetic Moment

An approximate method for describing the motion of charged particles in a magnetic field as a combination of cyclotron gyration and drift motion is called the adiabatic or **drift approximation**. In the adiabatic approximation certain quantities can be assumed to be conserved. These quantities are called **adiabatic invariants**. One of these is the angular momentum Mvr. The angular momentum of the cyclotron gyration is obtained if we substitute the cyclotron radius r_c for r:

$$Mvr_c = \frac{Mv^2}{\omega_c} = \frac{M^2v^2}{ZeB}.$$

Thus, for the cyclotron gyration the law of conservation of angular momentum is equivalent to constancy of the quantity

$$\frac{v^2}{B} = \text{const.}$$

Under constant conditions this relationship is exact; if there are small changes in the external conditions in the course of one gyration (adiabatic conditions), the relationship is approximate.

Let us multiply v^2/B by the constant factor M/2. Then

$$\frac{Mv^2}{2B} = \text{const}$$

or

$$\frac{E_\perp}{B} = \text{const,}$$

where $E_\perp$ is the kinetic energy of gyration in the plane perpendicular to the magnetic field. We will consequently refer to this as the "transverse energy." In adiabatic particle motion the transverse energy varies in proportion to the magnetic field. It is known that the magnetic moment μ in a magnetic field B has an energy given by

$$E_\perp = \mu B$$

proportional to the magnetic flux density.

Thus, the ratio $E_\perp/B$ is the magnetic moment of the orbit.

$$\frac{E_\perp}{B} = \mu.$$

The conservation of this quantity is called the adiabatic invariance of the magnetic moment: μ = const.

It is evident that the magnetic moment, determined in this way, does not pertain to the particle,* but to its orbit, i.e., the cyclotron circle on which it gyrates. In order to take advantage of the calculation of the magnetic moment, we will make an approximation in which we do not examine the particle motion, but rather, the motion of the center of the cyclotron circle. This is called the guiding-center approximation. The condition which must be satisfied if the guiding-center approximation is to be valid is the same as for the adiabatic or drift approximation:

$$\omega_c t \gg 1,$$

where ω_c is the cyclotron frequency and t is the characteristic time during which the external conditions change.

A careful mathematical analysis shows that the deviation from invariance of the magnetic moment is exponentially small, i.e., the correction term is of order $\exp(-\omega_c t)$. Therefore, for large values of the cyclotron frequency the constancy of the orbital angular momentum is maintained to a very high degree of accuracy.

The Adiabatic Traps

The invariance of the magnetic moment is very important in the theory of magnetic plasma containment. It presents the possibility of preventing particle loss along the magnetic field. A device that uses this principle is called an adiabatic magnetic trap. A schematic diagram of such a trap is shown in Fig. 15.

---
* Although it is sometimes erroneously called the "magnetic moment of the particle," it is more nearly correct to speak of the magnetic moment of the orbit.

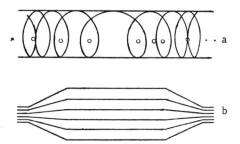

Fig. 15. An adiabatic magnetic trap. a) Schematic diagram of the windings; b) magnetic field lines.

The trap is a cylinder in which external windings establish a longitudinal magnetic field B_0. Near the ends of the cylinder the windings are more closely spaced (or a higher current is used), thus establishing regions of stronger field, which we shall call **magnetic plugs** or **magnetic mirrors**. In these regions the field has the value RB_0. The coefficient R is called the **mirror ratio** (R > 1).

Consider a charged particle gyrating in a cyclotron orbit with transverse velocity $v_\perp$ in the magnetic field B_0 near the center. As the guiding centers of these orbits move along the magnetic field lines into the region of higher field RB_0, by virtue of the conservation of magnetic moment, the transverse velocity must increase by a factor $\sqrt{R}$. However, the total energy

$$M(v_\parallel^2 + v_\perp^2) = \text{const},$$

must also be conserved (here $v_\parallel$ is the longitudinal velocity of the guiding centers along the field lines). From the conservation of energy it follows that the increase in the transverse velocity must result in a reduction of the longitudinal velocity.* If we denote the velocities in the region of increased field by primes,

$$v'_\perp = \sqrt{R}\, v_\perp,$$
$$v'^2_\perp + v'^2_\parallel = v_\perp^2 + v_\parallel^2;$$

---

* The reduction in longitudinal velocity takes place just as though the orbital magnetic moment were subject to a force $F = -\mu \nabla B$.

thus,

$$v'^2_\parallel = v^2_\parallel - v^2_\perp (R-1).$$

If the particle has a very large transverse velocity, the longitudinal velocity goes to zero in the mirror region. Equating the expression for $v'_\parallel$ to zero we obtain the conditon:

$$v_\perp = \frac{v_\parallel}{\sqrt{R-1}}.$$

Particles, with transverse velocity greater than this critical value will be reflected from the regions of higher field (hence, the name "mirror"). In other words, this region prevents particles from escaping from the trap (hence, the name "plug").

The same conclusions can be drawn from a geometrical approach: for particle motion along a helical path the velocity vector gyrates (precesses) around the magnetic field line at a constant angle θ given by

$$\tan\theta = \frac{v_\perp}{v_\parallel}.$$

As we have seen, all particles are trapped for which

$$\tan\theta \geqslant \frac{1}{\sqrt{R-1}}.$$

However, the trap cannot confine particles which have velocity vectors lying inside the cone described by

$$\tan\theta < \frac{1}{\sqrt{R-1}}.$$

This is the so-called **loss cone**. The cone becomes narrower as the mirror ratio R increases. If they do not experience collisions inside the trap all particles with directed velocities lying inside the escape cone will be lost; particles with velocities in other directions will remain trapped. Actually, collisions between particles change the velocity directions and throw particles into and out of the loss cone. As a result there is a continuous leakage of particles from the trap; the escape velocity depends on the collision frequency.

As we have seen, a magnetic trap only confines particles with sufficiently large transverse velocities, i.e., hot plasmas. Heating of a plasma inside the trap is not desirable: heating increases the leakage greatly. Therefore the method of injection is used to fill a trap with plasma. Fast ions are injected into the trap from a separate ion source (see below). A number of difficulties also arise in this case.

First of all, to establish a neutral plasma it is necessary to compensate for the ion space charge with electrons injected from outside. The moving electrons experience a force due to the space charge electric field. Large traps require large electron currents and these are not easily available.

Second, if the magnetic field does not allow particles to escape from the trap, then it also makes it difficult for them to enter. A compensating coil of appropriate configuration is used to establish a magnetic channel through which the injection can take place.

Third, the laws of mechanics, which govern the adiabatic motion of particles, indicate that in describing closed trajectories the particles must return to their starting point and strike the injector, causing them to be lost from the trap. In order to avoid this difficulty it is necessary to make the particle motion somewhat nonadiabatic. In the injection process we can increase the magnetic field rapidly. The rate of rise must be such that the field changes significantly in one cyclotron orbit. Technically this is rather difficult. A more practical solution is to inject molecular ions D_2^+ which, after entering the trap, dissociate into atomic ions and neutral atoms. The collision process is highly nonadiabatic; the atomic ions have a cyclotron radius which is only half that of the molecular ions. Therefore, after dissociation the particles do not gyrate back into the injector. This method of injecting molecular ions is used in the Soviet Ogra experiment and in the American DCX.

Drift in an Inhomogeneous Field

Let us now try to understand how particles move in an inhomogeneous magnetic field.

The inhomogeneity can be parallel or transverse to the direction of the field. A transverse inhomogeniety means a compression or rarefaction of field lines (Fig. 16). A longitudinal inhomogeniety means curved field lines (Fig. 17). If we assume the particle to be located at the center of its cyclotron circle, moving along the field lines (drift approximation), as a result of a longitudinal inhomogeniety the particle will experience a centrifugal force which gives rise to the **centrifugal drift**.

A transverse inhomogeniety means that the radius of the orbit in the strong-field region is smaller than in the weak-field region. Thus, the cyclotron circle acts as though it were stretched out across the field by a force which is proportional to the field **gradient** (i.e., the change in the magnetic flux density per unit length). This force produces a **gradient drift**.

If R is the radius of curvature of the magnetic field line a particle moving along the field line with a velocity v experiences a centrifugal force

$$F_c = \frac{Mv_\parallel^2}{R}$$

in the direction of the radius of curvature. This force produces a centrifugal drift with velocity

$$v_c = \frac{Mv_\parallel^2}{ZeBR}$$

in a direction perpendicular to both the field line and to the radius of curvature. If the magnetic field varies at right angles to its direction over a unit length by an amount $dB/dx \equiv \nabla_x B$ (this quantity is called the magnetic field gradient), the magnetic moment μ is subject to a force

$$F_B = -\mu \nabla_x B,$$

which causes a gradient drift with velocity

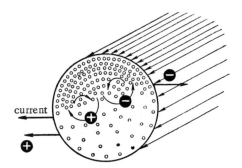

Fig. 16. Gradient drift. The magnetic field is directed into the page and increases toward the top of the figure. A positive particle drifts to the left, a negative particle to the right. The drift current is to the left.

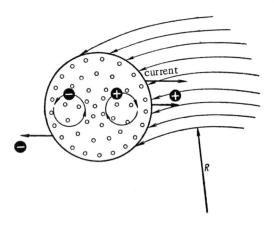

Fig. 17. Centrifugal drift. The magnetic field is directed into the page and is curved as shown. A positive particle drifts to the right, a negative particle to the left. The drift current is to the right.

DRIFT IN AN INHOMOGENEOUS FIELD

$$v_B = \frac{\mu}{ZeB} \nabla_x B$$

perpendicular both to the direction of the field and the direction of variation. Writing an expression for the magnetic moment of the cyclotron orbit

$$\mu = \frac{Mv_\perp^2}{2B},$$

we obtain the velocity of the gradient drift

$$v_B = \frac{Mv_\perp^2}{2ZeB^2} \nabla_x B.$$

The minus sign in the expression for F_B shows that force acts in the direction of decreasing magnetic field. If we are interested in the direction of drift as well as the magnitude of the velocity, these relations can be written in vector form:

$$\mathbf{v}_c = \frac{Mv_\parallel^2}{ZeB^2R^2} [\mathbf{R} \times \mathbf{B}];$$

$$\mathbf{v}_B = \frac{Mv_\perp^2}{2ZeB^3} [\mathbf{B} \times \nabla B].$$

Using these equations we can determine the correct sign of the motion.

All the expressions for the drift velocity which we have written apply both to ions and to electrons, for which we take $Z = -1$. Thus, the velocity of the centrifugal and gradient drifts for electrons can be written

$$\mathbf{v}_c = -\frac{mv_\parallel^2}{eB^2R^2} [\mathbf{R} \times \mathbf{B}],$$

$$\mathbf{v}_B = -\frac{mv_\perp^2}{2eB^3} [\mathbf{B} \times \nabla B].$$

The minus sign indicates that the electrons and ions move in opposite directions. An inhomogeneous magnetic field exerts a non-electrical force on the plasma and causes drift currents in a direction perpendicular to the field. The drift current carried by particles with charge Ze and mass M is given by j = nZev. Thus,

$$|j_c| = \frac{nMv_\parallel^2}{BR}$$

for the centrifugal drift and

$$|j_B| = \frac{nMv_\perp^2}{2B^2} \nabla_x B$$

for the gradient drift.

The total current is obtained by summing these expressions over all particles. The expression for the drift current can be put into a convenient form if we introduce the longitudinal and transverse plasma pressures, $p_\parallel$ and $p_\perp$. In the longitudinal direction the particles have one degree of freedom, hence

$$p_\parallel = nT_\parallel = nMv_\parallel^2.$$

In the transverse direction the particles have two degrees of freedom, therefore

$$p_\perp = nT_\perp = n\frac{Mv_\perp^2}{2}.$$

Finally, the drift current can be written

$$|j_c| = \frac{p_\parallel}{B} \cdot \frac{1}{R},$$

$$|j_B| = \frac{p_\perp}{B^2} \cdot \frac{dB}{dx} = c\frac{p_\perp}{BL},$$

where $L = B/\nabla_x B = dx/d\ln B$ is the characteristic length over which the magnetic field changes.

Drift in an inhomogeneous field complicates the problem of confining a plasma in a toroidal trap. If we imagine a torus in the horizontal plane, then both the centrifugal and gradient drifts produce vertical drift currents which cause charge separation (Fig. 18). As a result the plasma becomes polarized, i.e., a vertical electric field is created. The electric drift due to the presence of this field drives the plasma toward the outer wall.

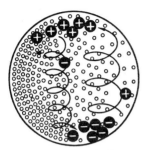

Fig. 18. Drift and plasma polarization in a toroidal trap.

Using the conducting-fluid model we obtain substantially the same result: the plasma must move outward. However the analysis based on the drift concept is more accurate and also shows a way to overcome these difficulties. If we produce a helical field then the charge separation is eliminated by the current which flows along the magnetic field lines since any given field line now moves from the top to the bottom.

Polarization Drift

We have investigated three basic forms of drift motion: electric drift, centrifugal drift, and gradient drift. These are described completely by the motion of particles in time-independent electric and magnetic fields which vary slowly in space. In addition there are possible drifts caused by extraneous nonelectrical forces, e.g., the force of gravity. All of these excite drift currents. If the particles experience a constant or slowly varying acceleration, then motion will result as though under the action of an inertial force equal to the product of the mass and acceleration. This force gives rise to an inertial drift and to a corresponding drift current. An especially important case, in which acceleration results from a change in the rate of electrical drift, is caused by variations in the electric field. This particular case of inertial drift is therefore called the polarization drift. This name is significant since the cause of the acceleration is the change in the electric field inside the plasma. This usually results, not so much from an external field, but from polarization of the plasma due to charge separation within the plasma itself.

The polarization drift is distinguished by the following unique characteristics. Assume that the electric field is directed at right angles to the magnetic field. This causes a drift motion of all particles, regardless of the sign of the charge, in the same direction (perpendicular to both the electric field and the magnetic field). If the electric field is an alternating field then an inertial force arises which is independent of a charge, acting in one direction. Referring to the general law for drift motion derived earlier we see that this force causes the electrons and ions to move against each other in a direction perpendicular to the inertial force and the magnetic field, i.e., parallel to the electric field. Therefore, polarization drift gives rise to a current along the electric field if it is perpendicular to the magnetic field. This drift current is superimposed on the transverse conductivity of the plasma and very often dominates it.

If the rate of change of the electric field is $\dot{E}$, then the electric drift will take place with an acceleration

$$|\dot{v}_E| = \frac{\dot{E}}{B},$$

which corresponds to an inertial force

$$F = M \frac{\dot{E}}{B}.$$

This force gives rise to a polarization drift with velocity

$$|v_p| = \frac{M\dot{E}}{ZeB^2}.$$

For the electrons this force is smaller than for ions by a factor of Zm/M, i.e., a factor of several thousand; hence in examining the polarization drift we can neglect the motion of the electrons. The drift current, carried by the ions is

$$|j_p| = n_i Ze |v_p| = \frac{Mn_i}{B^2}\dot{E} = \frac{\rho}{B^2}\dot{E},$$

where $\rho = Mn_i$ is the plasma mass density.

The current density is therefore proportional to the rate of change of the electric field. This means that the current associated with the polarization drift possesses properties similar to those of a d i s p l a c e m e n t   c u r r e n t which arises in polarization of a dielectric. This feature justifies the name "polarization drift." In accordance with the laws of electrodynamics governing a medium with dielectric constant ε, the displacement current is given by

$$|j| = \varepsilon_0 \varepsilon \dot{E}.$$

Therefore, in a changing electric field directed perpendicular to the magnetic field, the plasma can be described as a medium with dielectric constant

$$\varepsilon_\perp = \frac{\rho}{\varepsilon_0 B^2}.$$

The corresponding index of refraction is

$$n_\perp = \sqrt{\varepsilon_\perp} = \frac{\sqrt{\rho/\varepsilon_0}}{B}.$$

This result is correct only if the frequency of the changing field is small compared with the ion cyclotron frequency; otherwise it is not valid to use the drift approximation. Near the cyclotron frequency a dispersive dielectric constant is observed, i.e., the quantity ε depends on the frequency.

When we discuss plasma oscillations in a magnetic field we will see that the expression we have just obtained for the index of refraction applies to a wave that propagates with Alfvén speed. This is a low-frequency wave with an electric field which is perpendicular to the magnetic field. If the electric field is directed along the magnetic field the dielectric constant is altogether different. This result means that a plasma in a magnetic field is an anisotropic medium whose dielectric constant is a tensor quantity.

Rotating Plasmas

A plasma located in crossed magnetic and electric fields moves in the direction perpendicular to both of these fields. We now examine the interesting case in which the magnetic field is directed along the axis of a plasma cylinder and the electric field is radial. In this case the plasma velocity vector will be azimuthal, i.e., the plasma cylinder must rotate.

To establish a radial electric field we can place a conductor along the axis of the cylinder and apply a potential difference between this conductor and the walls of the cylinder. The axial conductor can also be replaced by a thin plasma jet shot from a plasma gun along the axis of the cylinder.

Rotating plasmas are of interest from several points of view. For example, theory shows that plasma rotation can facilitate stable plasma confinement. On this basis plasma traps incorporating rotating plasmas have been built, notably the Ixion and Homopolar experiments.

In the absence of hydrodynamic drag the azimuthal velocity of a rotating plasma approaches the electric drift velocity. In this case a sizable quantity of energy can be stored as kinetic energy of rotation (as in a flywheel).

A plasma cylinder with a radial electric field can also be regarded as a cylindrical capacitor. We have seen that a plasma can have an enormous dielectric constant. The energy associated with the electric field in a capacitor is proportional to the permittivity of the dielectric material. The energy of the plasma cylinder can be correctly regarded either as the electrical energy of a plasma capacitor or as the kinetic energy of the rotating plasma. A simple calculation shows that the numerical value of the energy is the same from either point of view.

The Magnetization Current

In a magnetoplasma the thermal motion of particles at right angles to the magnetic field is a cyclotron gyration. Each of the rotating charged particles forms a circular current in the plane perpendicular to the magnetic field. These current loops possess magnetic moments, which is why they are called **magnetization currents**. The ions and electrons rotate in opposite directions, but the currents they carry are additive. The direction of the current is commonly taken to be the direction of motion of the positive charges.

The magnetization current is excited by the external magnetic field. However, the field set up by these currents is opposite in direction to the external field, thereby weakening the latter. In the absence of other currents the plasma behaves as a **diamag-**

THE MAGNETIZATION CURRENT

netic medium. For this reason the magnetization current is also known as the diamagnetic current. In addition to these currents, drift currents and conduction currents can also flow in the plasma. The magnetic fields of these currents are oriented in various ways. For example, a conduction current can, in certain cases, establish a magnetic field which, because of collisional anisotropy, will be directed parallel to the external field rather than antiparallel, as is the case with a diamagnetic field. In this case the plasma acts as a paramagnet.

In a homogeneous plasma the magnetization current is not in evidence except as an implicit source of the magnetic moment. The reason is shown clearly in Fig. 19a. At each point we can imagine the upper part of one cyclotron circle to be in contact with the lower part of another. The current flows in opposite directions in the upper and lower parts of the circle. The total magnetization current in a homogeneous plasma is therefore equal to zero. A different picture arises, however, if the density of cyclotron circles or their radii are distributed nonuniformly in space (Fig. 19b). In this case the currents flowing in the upper and lower parts of the cyclotron circles do not cancel. The difference, i.e., the nonvanishing part, gives a finite magnetization current. The density of the cyclotron circles is just the plasma density and their radii depend on the plasma temperature and magnetic flux density. If any or all of these quantities vary in the direction perpendicular to the magnetic field, a magnetization current appears in the plásma. It will be perpendicular both to the magnetic field and to the direction of the gradient. A nonvanishing magnetization current is therefore a basic property of an inhomogeneous plasma.

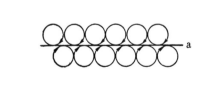
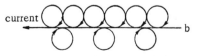

Fig. 19. Origin of the magnetization current. a) Homogeneous plasma; b) inhomogeneous plasma.

If we calculate the current flowing in the upper and lower halves of the cyclotron circles, we can find the total magnetization current density

$$|J_M| = \frac{dn\mu}{dx},$$

where n is the particle density, μ is the orbital magnetic moment, and x is the distance perpendicular to the field in the direction in which the quantity $n\mu$ varies. The magnetization current is perpendicular both to the direction of the magnetic field and to the x-direction. In vector form

$$\mathbf{j}_M = -\nabla_x \, n\mu,$$

where μ is the vector magnetic moment, which is perpendicular to the plane of the cyclotron circles in the direction of the diamagnetic field. Then, as we have just noted, the magnetization current will be carried by particles of one species with a fixed value of the orbital magnetic moment. In order to obtain the total magnetization current it is necessary to sum the currents carried by particles of all types. If we use the expression

$$\mu = \frac{Mv_\perp^2}{2B}$$

and assume that the plasma is in thermal equilibrium so that

$$\frac{Mv_\perp^2}{2} = T_\perp,$$

then

$$n\mu = \frac{nT_\perp}{B} = \frac{p_\perp}{B}.$$

We add the pressure contributions of the various particle species. Thus, for a thermal inhomogeneous plasma the total magnetization current is

$$|J_M| = \frac{d}{dx}\left(\frac{p_\perp}{B}\right)$$

or, in vector form,

$$\mathbf{j}_M = -\nabla_x \, \frac{p_\perp \mathbf{B}}{B^2}.$$

If the magnetic field is directed out of the page and the quantity $p_\perp/B$ (or $n\mu$) increases upward, then the magnetization current will point to the left (see Fig. 19). If the quantities dB/dx and

$d/dx\,(p_\perp/B)$ have the same sign, then the current associated with the gradient drift and the magnetization are in the same direction.

The Quasi-hydrodynamic Approximation

In the conducting-fluid model a number of forces which act on the plasma are related to the pressure force. In an ordinary gas of neutral particles pressure transfer takes place solely by means of collisions. Thus, the pressure force acts only at densities high enough so that frequent collisions occur. An approximate method for describing equilibrium and gas motion in terms of a pressure is called the hydrodynamic approximation. The hydrodynamic approximation can also be applied to dense plasmas just as for any dense gas. However, it turns out that even in tenuous plasmas in which collisions do not play a significant role, one can imagine a pressure that acts on each particle at right angles to the magnetic field. The physical mechanism for pressure transfer is however, quite different from that in the dense-plasma situation. The mechanism in question is associated with the interaction of the drift current and the magnetization current.

A description of the behavior of tenuous plasmas in terms of a pressure is called the quasi-hydrodynamic approximation. The applicability condition is the same as for the drift or adiabatic approximation.

Let us now seek to explain quantitatively the mechanism for pressure transfer in a rarefied plasma. We denote the perpendicular plasma pressure by $p_\perp$. The magnetic flux density B varies in the x-direction, perpendicular to the field B. Then the total current due to the gradient drift current and the magnitization current is

$$|j| = |j_B + j_M| = \left[\frac{p_\perp}{B^2}\cdot\frac{dB}{dx} + \frac{d}{dx}\left(\frac{p_\perp}{B}\right)\right].$$

This total current density is perpendicular both to B and x. Performing the indicated differentiation and clearing terms we find

$$|j| = \frac{1}{B}\frac{dp_\perp}{dx}.$$

If we recall the general formula for the drift current

$$|j_D| = n \frac{F_\perp}{B},$$

we see that the total transverse current can be accurately expressed as a drift current under the action of the force:

$$F_\perp = -\frac{1}{n} \cdot \frac{dp_\perp}{dx}.$$

In this way we can introduce a transverse pressure in place of the gradient drift and the diamagnetic current. We find an expression which is hydrodynamic in appearance. If we are interested in the direction of the current, we must rewrite this expression in vector notation

$$\mathbf{j} = -\frac{1}{B^2}[\nabla p \times \mathbf{B}].$$

The condition for magnetostatic equilibrium of a plasma perpendicular to the magnetic field can now be derived from Maxwell's equation (neglecting the displacement current):

$$-\frac{dB}{dx} = \mu_0 j.$$

Substituting the expression for the current, we obtain

$$-\frac{dB}{dx} = \frac{\mu_0}{B} \cdot \frac{dp_\perp}{dx},$$

which, after integration, tells us that the sum of the magnetic and kinetic pressures is a constant:

$$\frac{B^2}{8\pi} + p_\perp = \text{const.}$$

In an inhomogeneous plasma, a current must flow at right angles to the magnetic field. The existence of this current is a result of both the drift in the inhomogeneous field and the inhomogeneity of the cyclotron gyration (magnetization current). If the magnetic field only varies at right angles to itself, then the gradient drift and the magnetization current can be combined. The effect is that of a drift current caused by the pressure force. This current is often quite significant in a plasma located in a strong magnetic field. We are accustomed to thinking that an electric field causes a current and a pressure gives rise to a mass motion. In a plasma we find that the situation is quite different. We have seen that an electric field can give rise to drift motion of the plasma as a whole. We now see that a pressure can excite drift currents in a plasma.

Hydromagnetic Plasma Instabilities

The further we pursue our study of the plasma state, the clearer it becomes that instabilities play a central role in plasma phenomenon. In very many cases the equilibrium of a plasma is stationary,* i.e., in the absence of a perturbation the plasma state can persist indefinitely, although it is unstable to an arbitrarily small perturbation.

In order to understand the nature of a simple type of instability we shall illustrate it by the example of a system which is simpler and more familiar than a plasma. Into a test tube containing a small amount of water we carefully pour some sulfuric acid. If this is done very accurately a two-layer system is obtained in the test tube (Fig. 20a). The layer of heavy liquid lies on top of the lighter (water) layer. This system is in equilibrium. The force of gravity which presses the heavy upper layer down is balanced by the pressure of the lower layer of light fluid. However, this equilibrium is not favorable from an energy point of view. If the lower liquid could change places with the sulfuric acid, leaving the latter on the bottom and the water on top, the potential energy of the system would decrease. The energy liberated in this process is converted into kinetic energy, which is used in

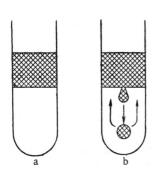

Fig. 20. A two-layer system (a) and its instability (b).

* By "stationary," we mean a condition which does not change with time unless it is subjected to a perturbation. Stationarity should not be confused with stability.

the process in which the interchange occurs. This process is energetically favorable but the direct actions of the forces responsible for the exchange do not tell the whole story.

The force of gravity can only cause vertical motion; on the other hand, the exchange of positions of the heavy and light layers means that the lower layer must somehow make room for the upper one, i.e., some sort of horizontal motion is necessary. Thus, we say that the heavy fluid layer lying on the light layer is in equilibrium, but that the equilibrium is unstable with respect to small perturbations. If even one small drop of the heavy liquid should happen to drop (this is called a perturbation) it leaves room in the upper layer, which the light fluid can then fill. This movement, in turn, frees space in the lower region which can be filled by heavy fluid, etc. (Fig. 20b). The subsequent motion consists of counterstreaming of the two liquids. Naturally, they cannot flow through each other at the same point; therefore the motion takes on a circulatory character. At one point the heavy fluid moves down and at another the light fluid moves up. This motion is called convection. Convective motion cannot begin without a perturbation because in equilibrium no location is distinguishable from another and a priori, there is no way of determining at what points the fluid should rise or fall. The choice is determined by the initial arbitrarily small perturbation. As a simple example of an unstable equilibrium we might consider the famous ass postulated by the medieval scholar Buridan in the controversy concerning free will. At equal distances from the muzzle of this unfortunate animal there are two exactly identical bunches of hay. The poor ass, in the opinion of Buridan, must die of hunger, not knowing which bunch to choose. The error in Buridan's reasoning is that he did not anticipate small perturbations. The ass is actually in equilibrium, not knowing which choice to make. However, any random motion of the air carrying the smell of the hay, say, in the right-hand bunch, allows the choice to be made.

In the experiment with the sulfuric acid and the water we are only able to observe the beginning of the convection process. The mixing of the two liquids leads to a solution of sulfuric acid and water, accompanied by the appearance of heat; the further developments are not of interest to us. If we take two immiscible liquids, e.g., mercury and water, and attempt to float the mercury on the water (which is very difficult because of the large difference in their specific weights), the convection occurs very rapidly and the liquids change places completely. Stationary convective motion can also be obtained. This can be done by heating a layer of gas or liquid from below. As the heated material expands upward we obtain a light layer which will seek to change places with the heavy upper layer as in the preceding example. If the layer is heated from below, and the upper layer is cooled, then the circulatory convective motion will transport the heat continuously. Stationary thermal convection is very easily observed. Meteorologist observe it, for example, in the Sahara Desert where the solar rays pass freely through the clear air and heat the surface of the sand and the sand heats the air from below. Sometimes the process is called the Sahara effect. This process is easily simulated in the laboratory. It can be shown that for moderate velocities the layer breaks up into uniform hexagonal cells (Fig. 21); heated air moves up and cold air moves down in alternate cells. This is ordered, or laminar, convection. With increasing thermal flux and circulation velocity the motion becomes irregular and turbulent.

In plasmas or highly conducting liquids a new convective instability is possible; this is called the magnetohydrodynamic or hydromagnetic instability. Cells characteristic of this instability appear when a highly conducting plasma, inside of which there is no magnetic field, forms a boundary with spaces in which there is a magnetic field. In a thin layer near the surface of the plasma the magnetic field falls to zero abruptly from a finite value in the external space. Consequently, this surface layer contains surface currents. As we have seen earlier, such surface current layers are called skin layers and the plasma which they bound is known as a plasma with a "skin" (diamagnetic). If there is no magnetic field inside the plasma whatsoever, then the plasma is fully diamagnetic and the plasma and the external magnetic field can be regarded as two immiscible liquids. If the combined energy configuration is not favorable a convective instability can arise. The force, which causes the convective motion, will not be the force of gravity in

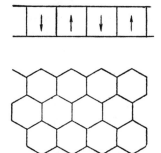

Fig. 21. Convective cells.

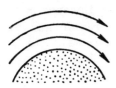

Fig. 22. A convex plasma boundary.

this case, but rather a pure magnetic force, the tension of the magnetic field lines. The field lines, similar to strings, tend to shorten, i.e., to occupy a position in which the lengths are minimized. Assume that a fully diamagnetic plasma is separated from the magnetic field by a convex surface (Fig. 22). In other words, the center of curvature of the surface (and the magnetic field lines) is inside the plasma. This configuration is not favorable energetically because the magnetic field lines are stretched. Now imagine that the magnetic field and the plasma change places. The field lines contract, the magnetic field energy decreases, and the plasma energy remains unchanged. The new state is energetically more favorable. Therefore, a fully diamagnetic plasma with a convex surface and no frozen-in magnetic field is unstable. The instability is convective and is essentially the same as the instability of a heavy fluid floating on a light one.

The same phenomena can be described in other terms. We say that a plasma in which there is no magnetic field is diamagnetic, i.e., it tends to move in the direction of decreasing external magnetic field. If the field lines are convex outward, then their density, i.e., the magnetic flux density, decreases in the same direction. Thus, a fully diamagnetic configuration for which the magnetic field decreases with distance from the plasma, is unstable.

The discussion of instabilities given above has been based on the energy principle, which says that a system tends to minimum potential energy. We can arrive at the same answers by a perturbation approach in which we examine the motion in a system subject to various perturbations. It can be shown that in an ideal plasma (no dissipative processes) any perturbation of a stable equilibrium state results in simple harmonic oscillations of the plasma surface, similar to waves on water. Finite conductivity,

Fig. 23. A transverse perturbation. a) The ends of the field lines are free; b) the ends of the field lines are fixed.

Fig. 24. Longitudinal perturbation.

viscosity, and other dissipative processes tend to damp the oscillations. In an unstable plasma the nature of the motion depends on the form of the initial perturbation. If the perturbation extends across the field lines (Fig. 23) the same surface oscillations are excited as in the stable case. If the perturbations extend longitudinally along the field lines (Fig. 24), the field lines are pushed apart and an unbounded aperiodic growth of the perturbation results. The growth rate of the perturbation is larger, the smaller the wavelength. Short wavelength perturbations ("ripples" on the plasma surface) grow rapidly, without bound. For shortwave perturbations the growth rate is inversely proportional to the square root of the wavelength.

Pinch Instability

A simple and thoroughly investigated hydromagnetic instability is the instability discovered in the course of studying a plasma column (pinch) compressed by the magnetic field produced by current flowing through the column itself. The pinch instability is essentially the same as the instability of any flexible conductor along which current flows. There is a circular magnetic field around the straight conductor; the strength of this field is inversely proportional to the radial distance from the axis. The field lines are everywhere convex outward and their center of curvature lies on the axis, i.e., inside the plasma. Therefore, it follows from the energy principle that if there is no magnetic field inside the plasma (completely diamagnetic) the pinch must be unstable with respect to convective motion. It is possible to give a clear picture of the instability.

Assume that we have a small perturbation which causes a slight bending of the column. On the inside of the bend the field lines are crowded together; outside they are separated (Fig. 25). As a result the magnetic flux density (and therefore the magnetic pressure) is increased on the inside of the bend and causes the column to bend still further. A detailed calculation shows that deformations of various kinds can arise. In order to classify these

Fig. 25. Pinch instability caused by magnetic pressure.

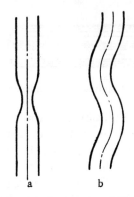
Fig. 26. Sausage instability (a) and kink instability (b).

deformations it is convenient to use the azimuthal number m which indicates how many times the direction of the deformation changes sign as one goes around the pinch. The number m = 0 denotes the "sausage" deformation instability, m = 1 denotes the kink instability, etc. (Fig. 26).

Stabilization by Frozen-in Magnetic Fields

If a magnetic field is frozen into the plasma the plasma can be stabilized, i.e., certain instabilities can be eliminated. However, stabilization requires that the magnetic field lines must not be closed inside the plasma, but must intersect solid metal conductors. In the case of a long pinch these conductors are the electrodes which carry the plasma current.

The mechanism by which frozen-in magnetic fields stabilize a plasma is very simple. If the field is truly frozen, any motion of the plasma perpendicular to the field lines causes them to be bent (see Fig. 23). But bending of the field lines causes a stretching, which is resisted by the elasticity of the lines (stretching requires energy). If the ends of the field lines are fixed in solid conducting surfaces, their elasticity inhibits the deformation and the field lines confine the plasma which adheres to them. The field lines contribute a stiffness to the plasma in the way iron rods are used to reinforce concrete.

We should emphasize that the interior magnetic field stabilizes the plasma only so long as the field is truly frozen to the plasma, i.e., the plasma conductivity must be very high. In weakly ionized plasmas with low conductivity the opposite effect is possible: the magnetic field can cause an instability of the current which flows along this field in the plasma. Such an instability gives rise to a pondermotive force which acts on any curvature of the currents which disturbs the parallelism of the field; the instability tends to orient the bend in a direction perpendicular to the field. Thus, in weakly ionized plasmas that carry currents there is a critical magnetic field, beyond which there is an anomalous diffusion of plasma perpendicular to the magnetic field.

Even in a perfectly conducting plasma the frozen-in magnetic field does not give complete stabilization with respect to all possible perturbations. In the case of the linear pinch optimum stabilization is obtained when there is only a longitudinal field inside the plasma and the current only flows at the surface. In other words, the longitudinal field is concentrated inside the plasma and the circular field is outside it; the fields are spatially separated. For such separated fields, the pinch is stabilized with respect to all perturbations with azimuthal numbers m, not equal to unity, if it is possible to provide a longitudinal field that strong in comparison with the circular field established by the current itself. For stabilization against the sausage instability (m = 0) the interior field must be at least $1/\sqrt{2} = 0.707$ times stronger than the external circular field. Perturbations with azimuthal numbers m = 2 and higher are even more easily stabilized with separated fields. However, kink perturbations (m = 1) can only be stabilized by the frozen-in magnetic field for short wavelengths. Pinch stabilization with respect to longwave kinks requires the use of another method, namely, locating the pinch in a chamber with nearly highly conducting walls. If the conductivities of the plasma and the walls are both high, the magnetic field cannot penetrate either and forms an elastic cushion which prevents the plasma from coming in contact with the wall.

Therefore, the combination of a frozen-in longitudinal field inside a plasma and a conducting wall outside can, in principle, be used to stabilize a highly conducting pinch with separated fields. However, the complete separation of the longitudinal and circular fields required for stabilization can only exist for a time short compared with the skin diffusion time. If the current does not only flow at the surface, but is also distributed inside the plasma, the fields are no longer separate. Magnetic field diffusion due to finite conductivity leads to mixing of the fields and the combination of longitudinal and circular fields forms a helical field. In the presence of a helical field a helical perturbation with the same pitch as the field cannot be stabilized simply by increasing the field strength. In order to stabilize the plasma with respect to such perturbations it is necessary to "cross" the magnetic field lines in such a way that their elasticity resists any perturbations. In a pinch this can be done by applying an external field in the direction opposite to that of the internal field. A considerably greater range of stabilization possibilities can be realized by using special helical field configurations. The stellarator configuration is an example of this kind of field.

Interchange or Flute Instabilities

A simple kind of hydromagnetic instability occurs when a plasma, with no internal field, comes in contact with a magnetic field in free space. If the strength of the magnetic field decreases with distance from the boundary of the plasma (in which case the field lines must be convex outward), then it is energetically favorable for the plasma to change places with the magnetic field. This phenomenon is called the interchange instability because the motion can be described as an interchange of the lines of force in space. In this case the field lines move as a whole, conserving their form and direction. The surface perturbation takes the shape of flutes along the magnetic field lines. Therefore, this instability is also known as a flute instability. This terminology provides a clear description of a plasma which separates the magnetic field lines and leaks into the space between them.

There are two simple ways in which flute instabilities can be stabilized. First of all, this instability requires that the field lines be free to move in space. If the ends of the lines are "frozen" to a solid conductor, the interchange is impossible. However, in this case the plasma must be in contact with the metal walls or electrodes. The second method for suppressing the flute instability in-

INTERCHANGE OR FLUTE INSTABILITIES

volves establishing a stabilizing magnetic field inside the plasma, directed at an angle to the external field (in the simplest it is perpendicular). Near the dividing surface the lines of force form a "woven mesh" (Fig. 27a). This mesh prevents the plasma from leaking out.

In examining flute instabilities we have assumed that there is a sharp boundary between the plasma and the vacuum. A sharp boundary can exist in a stationary state only if the plasma conductivity is essentially infinite. If we take account of the finite conductivity the boundary can only be assumed to be sharp for a time small in comparison with the skin diffusion time. Further diffusion of the magnetic field leads to smearing of the plasma boundary. The effect of this transition from a sharp boundary to a diffuse boundary provides an instructive example of how complex and delicate the criteria for plasma stability are.

If there is no stabilizing field inside the plasma an interchange between the plasma and the magnetic field lines is more favorable energetically, the sharper the boundary. In this case boundary smearing stabilizes the plasma. However, if a stabilizing field is present in the plasma and if it is perpendicular to the external field, boundary smearing produces the opposite effect. In this case magnetic field diffusion leads to a m i x i n g of the mutually perpendicular fields. In the diffuse boundary zone, a helical field (Fig. 27b) is produced rather than a "mesh." The plasma can then leak into the space between the helical field lines. The perturbation is in the form of a helical flute, directed along the helical field lines. Thus, boundary smearing increases the stability in a field-free plasma; on the other hand, in a plasma stabilized by an internal magnetic field boundary diffusions leads to instability.* This example teaches us how cautious we should be in analyzing the effect of various factors on the hydromagnetic

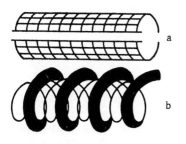

Fig. 27. Stabilization of plasma by two mutually perpendicular fields (a) and breakdown of stability due to field mixing (b).

*Strictly speaking, boundary smearing reduces the critical value of β (ratio of kinetic pressure to magnetic pressure) beyond which the plasma is unstable.

instability of plasmas. Each concrete case must be treated individually.

Flute instabilities are undesirable for plasma confinement in adiabatic traps with magnetic plugs (mirrors). In the transition region from the constant main field to the stronger mirror field, the field lines are unavoidably convex outward (see Fig. 15). In this region tongues of plasma must leak out between the flutes and expand along the field lines; ultimately the plasma strikes the walls. Experiment shows that the plasma in an adiabatic trap decays rather quickly. This rapid decay can be explained by flute instability. To prevent this it has been proposed to freeze the ends of the field lines in metal conductors (as in Fig. 23b).

There is also another proposal — to change the adiabatic trap into an opposed system, i.e., a trap with opposed fields. In this system the two coils carry current in opposite directions and establish the magnetic-field configuration shown in the right-hand part of Fig. 11. This configuration is similar to the magnetic field of a four-pole arrangement (quadrupole). As a consequence traps with opposed fields are sometimes called quadrupoles. Here the magnetic field is concave everywhere and flute instability cannot occur. Because of the presence of the "equatorial" cusp and the two cusps, at the poles, this trap is also known as a cusp. These cusps act like "holes," two at the poles and one along the equator. It has also been proposed to shoot a "blob" of plasma (called a plasmoid) with large ordered velocities into the trap. A plasma gun can be used for this purpose (see Fig. 11). After the energy of the ordered motion is transformed into thermal energy, the plasma in the trap expands and the particle velocities become smaller than that needed for rapid loss through the cusps.

Diffusion of Opposed Fields

When an alternating field is applied to a plasma a situation can sometimes arise in which there is a magnetic field of one sign frozen into the plasma, while the field outside has the opposite sign. Similarly, regions with opposite fields can come into contact for other reasons, e.g., randomly moving magnetic regions in the solar atmosphere. Finally, deliberate creation of an external field, opposite to that inside, has been proposed as a method for stabilization of pinches.

In all cases of contact between oppositely directed fields, interdiffusion of the fields leads to peculiar phenomena. Whereas the mixing of perpendicular fields, as discussed above, leads to the formation of a helical field, interpenetration of opposed fields leads to their annihilation. We will examine an instructive case, in which fields are equal in magnitude and opposite in sign. Interdiffusion leads to the formation of a neutral layer in which there is no magnetic field (Fig. 28). The magnetic pressure is equal to zero in the neutral layer whereas it is subject to magnetic pressure forces, from both sides, thus causing the layer to be compressed. The field-free matter in the neutral layer resists the compression only as an ordinary gas. Under compression the matter reaches a high density and temperature. If field diffusion takes place rapidly, the compression can lead to the formation of shock waves which converge from both sides of the layer and come together in the center. The waves reflected after the collision can generate plasma oscillations.

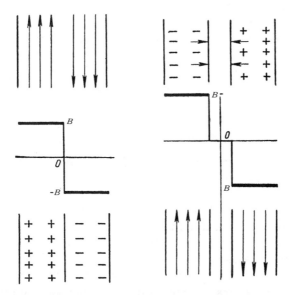

Fig. 28. Diffusion of opposed fields with the same strength and formation of a neutral layer.

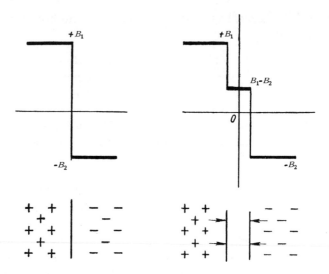

Fig. 29. Diffusion of opposed fields of different strengths with the formation of a partially neutralized layer.

As always, compression of a neutral layer can be described in terms of magnetic pressure as we have just done; however, we can also give a description in terms of pondermotive forces. In this case we say that mutual cancellation of the fields produces a steep magnetic field gradient, perpendicular to the original field direction; according to the laws of electrodynamics this gives rise to a strong current, perpendicular to both the field and the gradient. The interaction of this current with the magnetic fields produces a pondermotive force, which compresses the neutral layer.

If the fields are unequal in magnitude then a similar phenomenon is observed, but less marked. In this case the fields cancel incompletely. Instead of obtaining a neutral layer we obtain a layer with a partially neutralized field in which the strength of the field will be very much smaller than in the layers on both sides (Fig. 29). The compression of this layer is similar to that which occurs in the previous case except that the compression stops earlier and the temperature and density are not as high. The compression of plasma neutral layers by interdiffusion of opposed fields finds important application both in nature and in the laboratory. The highest temperatures obtained in a laboratory plasma are achieved by this

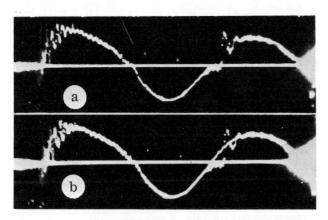

Fig. 30. Oscillograms of the field inside (a) and outside (b) the plasma in a θ-pinch.

method. This work was done in experiments with a θ-pinch, i.e., a plasma column which is compressed by a growing longitudinal magnetic field.

In practice the magnetic field cannot be made to increase monotonically; instead an alternating field is used. When the sign of the external field changes we obtain the pattern of mixing of opposed fields and the compression of a neutral layer described above. In this phase of the experiment very high temperatures are recorded (by several methods the temperature has been found to be $\sim$ 1 keV, i.e., of the order of 10 million degrees*); also neutron production is noted which is quite probably of thermonuclear origin. The plasma heating in this experiment is undoubtedly due to compression of the neutral layer. This can be confirmed from oscillograms of the field inside the plasma, as shown in Fig. 30 (curve a). For comparison, curve b shows the variation of the external field with time. At the time the external field changes sign the growth of the internal field slows down sharply. This is due to an expansion of the plasma together with the field frozen into it, which compresses the neutral layer. In the final stage of the layer compression a powerful shock wave is formed which rapidly compresses the plasma column. This is indicated in the oscillogram by the sudden increase in the frozen-in field. Multiple reflections

---

* When we speak of orders of magnitude, we are indicating the number to the nearest power of ten. Thus, even though 1 keV is 11.6 million degrees we say that the order of magnitude is 10 million degrees.

of the shock wave lead to pulsations of the plasma column which are seen on the oscillogram as pressure pulsations.

Completely anologous phenomena on a larger scale are observed as flares on the surface of the sun (known as c h r o m o - s p h e r i c   f l a r e s). Very accurate field measurements have been made using a s o l a r   m a g n e t o g r a p h; these show that the flares always occur at the boundary between two magnetoactive regions with opposite polarization (i.e., magnetic field direction).

The origin of the flares can be explained as follows. Deep in the solar interior convective motion produces magnetic tubes. The increased magnetic pressure causes the matter in the tubes to expand. Thus, the density inside the tubes becomes smaller than that of the surrounding plasma and the tubes float up to the surface of the sun, creating a magnetic field region there. If two tubes with opposite polarizations happen to be side by side, then the picture of opposed field mixing described above leads to the compression of a neutral layer and the formation of a converging shock wave. The heating of the compressed layer is accompanied by acceleration of some of the plasma material. The acceleration is produced both by density waves propagating outward through the plasma and by electromagnetic fields. These processes are responsible for all flare phenomena, e.g., the luminous flux and the ejection of c o r p u s c u l a r   s t r e a m s (i.e., streams of fast particles). When they reach the earth the streams cause polar auroras, magnetic storms, breakdown in radio communications, and bursts of intense cosmic rays. It is reasonable to assume that at least some of these effects are not caused by particles, but by magnetohydrodynamic shock waves which strike the earth after propagating through the tenuous magnetoplasma of interplanetary space.

Oscillations and Waves in Plasmas

Various kinds of oscillations and waves can be excited in a plasma. In other words a plasma has a large number of oscillatory degrees of freedom. As a rule, oscillatory processes which start in one place propagate in space from that point. A propagating oscillation of this kind is called a w a v e. Oscillatory processes are characterized by three basic quantities: the a m p l i t u d e, i.e., the height of the oscillation; the f r e q u e n c y, i.e., the number of oscillation cycles per unit time; and the p h a s e, i.e., the time at which the cycle passes through a characteristic point (maximum, minimum, etc.). The amplitude and the frequency are absolute quantities: They characterize a given oscillatory process as such and not in comparison with other processes. The phase is a relative quantity; it is arbitrary in itself and has meaning only in terms of the phase difference between two oscillatory processes.

Oscillatory processes in plasmas are associated with the simultaneous variation of a number of quantities which depend on each other. Some of these are primary and the others secondary. If the magnitude of an electric and magnetic field oscillates in a plasma, it can excite a periodic current. The current, interacting with the magnetic field, gives rise to a Lorentz (pondermotive) force which leads to motion in the plasma. Therefore, as a rule, any oscillatory process in a plasma is accompanied by a periodically varying velocity in the plasma which, in turn, leads to pressure oscillations. We can say that except in certain particular cases, plasma oscillations are simultaneously electromagnetic and hydrodynamic oscillations. However, in certain particular cases certain quantities dominate the oscillations and, speaking roughly, we can say that an oscillation is electrostatic, electromagnetic, magnetohydrodynamic, acoustic, etc. It should be kept in mind that these classifications are not universal. In a number of cases different kinds of oscillations are mixed and processes arise in which changes of all quantities characterizing the plasma are of equal importance.

The properties of oscillations are highly simplified when at small amplitudes. Such oscillations are called linear, since they are described by linear equations in the sense that terms containing quadratic quantities, derivatives of the amplitude, and higher orders of the amplitude can be neglected. The basic property of linear oscillations is that their frequencies are not amplitude dependent. The properties of nonlinear oscillations are very much more complex and our knowledge concerning plasma oscillations is confined largely to the linear regime.

Perturbations giving rise to oscillations at one point are propagated with velocities which are called the propagation velocity of the wave. In simple cases this velocity is independent of the frequency. However, in plasma physics we frequently encounter propagation velocities which are frequency dependent. This dependence is called **dispersion**. In the presence of dispersion we must differentiate between the **phase velocity** and **group velocity** in wave propagation. The phase velocity is that velocity with which a given phase is transported in space, e.g., the crest of the wave. All oscillatory quantities (velocity, field strength, etc.) moving with the phase velocity of the wave conserve their constant values. The variation of these quantities in time is compensated exactly by their spatial variation. The group velocity is the actual propagation velocity of the wave. Since the phase velocity does not correspond to an actual physical propagation, the limitation imposed on propagation speeds by the theory of relativity does not apply, i.e., the phase velocity can be larger than the speed of light. The wavelength is equal to the phase velocity multiplied by the period, i.e., divided by the frequency of the oscillation.

It is convenient to describe waves by means of a **wave vector**; the direction of this vector coincides with the direction of propagation and the length is equal to the **wave number**:

$$k = \frac{2\pi}{\lambda},$$

where λ is the wavelength. The phase velocity is

$$U_{ph} = \lambda f = \frac{\omega}{k}.$$

and the group velocity is

$$U_g = \frac{d\omega}{dk},$$

where ω is the angular frequency and f is the circular frequency:

$$\omega = 2\pi f.$$

In the absence of dispersion

$$U_{ph} = U_g = \text{const},$$

this follows because

$$\frac{d\omega}{dk} = \frac{\omega}{k}$$

for frequency-independent velocities.

The ratio of the velocity of light in vacuo to the phase velocity of the wave is the **index of refraction**

$$n = \frac{c}{U_{ph}} = \frac{kc}{\omega}.$$

In an unbounded space all perturbations excite **traveling waves**; the phase of these varies in space in correspondence with the phase velocity of propagation. In a bounded plasma, multiple reflections of the wave from the plasma boundaries can lead to the establishment of **standing waves** with phases which are everywhere the same. For this condition to obtain it is necessary that there be a **resonant** relationship between the phase velocity and the dimensions of the volume which encloses the plasma. Such a volume is called a **resonator** and oscillations of the standing wave type are called **characteristic oscillations** of the resonator. The frequency of these oscillations is the **characteristic frequency** of the resonator. The lowest of these characteristic frequencies is called the **fundamental frequency** and oscillations with higher frequencies are called **harmonics**. Intermediate between traveling waves in free space and standing waves in a resonator are waves which propagate along a channel (or tube), bounded by walls. Here the waves move along the axis of the channel and act as standing waves in the transverse cross section. These channels are known as **waveguides**.

If we imagine a swinging pendulum which is free to oscillate without being disturbed, the oscillation decays gradually because of friction and the oscillation energy is transformed into heat. This process is an example of **energy dissipation**. Plasma oscillations also decay because of various dissipative processes. The simplest mechanism for dissipation in plasmas is collisions between particles; these transform the energy of ordered oscillation into heat, i.e., into random particle motion. If an electric current flows in a plasma, as a result of the electrical resistance, i.e., finite plasma conductivity, part of the energy of the current is transformed into heat. This heat, liberated by the current, is called **Joule heat**. Joule dissipation is a special case of dissipation due to collisions since the electrical resistance of the plasma results from collisions between the current-carrying particles and other particles in the plasma. If the plasma moves as a whole, then collisions will manifest themselves as an internal friction; this is viscous dissipation. In a dense plasma this dissipation mechanism is dominant. In very tenuous plasmas the damping is independent of collisions. This anomalous dissipation is associated with the transformation of oscillation energy, not into heat, but into other kinds of oscillations or into general mass motion.

Electrostatic Plasma Oscillations

In the absence of a magnetic field, in addition to the usual acoustic oscillations there is only one other oscillation which is characteristic of plasmas. This oscillation is a result of charge separation. When we spoke of quasi neutrality above we noted that any displacement of charges of one sign in a plasma leads to oscillations. As noted previously, these oscillations are called electrostatic, or Langmuir, or simply plasma oscillations since they are a very general type of oscillation; indeed, the capacity for this kind of oscillation can be taken as part of the definition of a plasma. In a magnetic field, this kind of oscillation can be observed in pure form if the direction of the electric field and the motion of the particles are parallel to the magnetic field. In this case the magnetic field does not affect the motion of the particles.

In an ordinary gas any particle displacement leads to a pressure disturbance which is propagated at the speed of sound. In a plasma this process occurs only if particles with charges of both sign are simultaneously displaced, i.e., if no charge separation takes place. If particles of one sign are displaced with respect to particles of the other sign, in addition to the pressure disturbance there is also a space charge which produces an electric field.

Under the combined effect of the pressure force and the electric field there is a wave motion. If the plasma is cold and the electric field acts alone the oscillation frequency depends only on the plasma density. This characteristic frequency is called the plasma frequency. We have previously encountered this frequency when we discussed quasi neutrality and the time scale for charge separation. If we take the thermal pressure into account then dispersion appears and the plasma frequency is given by

$$\omega^2 = \omega_0^2 + k^2 U_s^2,$$

where $k = 2\pi/\lambda$ is the wave number (λ is the wavelength), ω is the angular frequency of the oscillation, ω_0 is the angular plasma frequency, and U_s is the sound speed. An equation of this type, which gives the frequency as a function of the wave number, is called a **dispersion relation**. Here the acoustic speed is taken to mean the propagation speed of a pressure disturbance caused by a perturbation of only those particles which participate in the charge separation. We usually concern outselves with **electron plasma oscillations** in which only the electrons are displaced. Thus, U is the **electron acoustic velocity**. This is found from the same formula as the ordinary acoustic velocity by substituting the temperature and density of the electrons.

Using the dispersion relation we find the phase velocity

$$U_{\text{ph}} = \frac{\omega}{k} = \sqrt{\frac{\omega_0^2}{k^2} + U^2}.$$

However, propagation does not occur for wavelengths smaller than the shielding distance h. The phase velocity of the electron oscillations is always greater than the acoustic speed. In the limit of very long wavelengths the frequency approaches the plasma frequency and the phase velocity $U_{\text{ph}} \to \infty$. This means that the entire plasma volume oscillates with a constant plasma frequency.

The direction of propagation of the plasma wave coincides with the direction of the electric field and the particle motion (as in an acoustic wave). Thus, the oscillations are said to be **longitudinal**, as opposed to **transverse** electromagnetic waves, in which the electric and magnetic fields lie in the plane perpendicular to the direction of propagation.

At high densities the plasma oscillations are very quickly damped by collisions. However, plasma oscillations are possible in a tenuous plasma in which sound cannot propagate because of the absence of a mechanism for pressure transport. In a tenuous plasma the role of the acoustic velocity is played by the quantity

$$U_s = \sqrt{3\frac{T_e}{m}},$$

which differs from the electron acoustic velocity only in that the adiabatic factor must be set equal to three (which corresponds to motion in one direction).

It is remarkable that in an arbitrarily tenuous plasma, regardless of collisions, the thermal motion leads to damping of plasma oscillations. This is known as Landau damping, an example of anomalous damping in a plasma. We will have more to say on this subject later.

The simplest and most important method for exciting plasma oscillations makes use of an electron beam. A beam of fast electrons, traveling through the plasma, causes displacement of the plasma electrons and excites plasma oscillations. If fast electrons appear as a result of some process in a plasma, they excite plasma oscillations. This phenomenon is one kind of oscillatory instability and is known as the two-stream instability.

Electrostatic Oscillations with Ions

In considering electrostatic oscillations we usually only examine high-frequency oscillations, in which case the ion motion can be neglected. We call these oscillations plasma oscillations. If we take the ion motion into account we find a plasma with no magnetic field can support two oscillation branches, both being associated with charge separation. One of these is a high-frequency oscillation; this is the electron plasma oscillation which has just been considered. The frequency, as we have just seen, starts at the electron plasma frequency and goes to higher frequencies. In a cold plasma the frequency approaches the electron plasma frequency. The inclusion of ion motion has a very weak effect on the electron branch. An accurate expression for the electron plasma frequency, including ion motion, is

$$\omega_0^2 = \frac{ne^2}{\varepsilon_0 m}\left(1 + \frac{Zm}{M}\right).$$

Noting that the electron mass m is very small compared with the ion mass M, we see that the ion correction can always be ignored. The dispersion relation for the electron branch (taking account of the ion motion) uses the sum of the squares of the electron and ion acoustic velocities rather than the electron acoustic velocity alone. However, since the ion acoustic velocity is always considerably smaller than the electron velocity the correction due to the ion motion is negligible.

The second, low-frequency, oscillation branch in a plasma oscillation (in the absence of a magnetic field) is due to the ion motion. The dispersion relation for this branch is

$$\omega^2 = \frac{\frac{mZ}{M}\omega_0^2 k^2 U_e^2 + \omega_0^2 k^2 U_i^2 + k^4 U_e^2 U_i^2}{\omega_0^2 + k^2(U_e^2 + U_i^2)}.$$

If we substitute the acoustic velocities

$$U_e^2 = \gamma_e \frac{T_e}{m}; \quad U_i^2 = \gamma_i \frac{T_i}{M},$$

where T_e and T_i are the electron and ion temperatures in energy units and γ_e and γ_i are the corresponding adiabatic coefficients (ratio of the specific heat at constant volume to the specific heat at constant pressure) the dispersion relation for the ion branch becomes

$$\omega^2 = \frac{k^2 \omega_0^2 \frac{Z\gamma_e T_e + \gamma_i T_i}{M} + k^4 \gamma_e \gamma_i \frac{T_e T_i}{mM}}{\omega_0^2 + k^2\left(\gamma_e \frac{T_e}{m} + \gamma_i \frac{T_i}{M}\right)}.$$

We can vary the wave number for constant ion and electron temperatures. Then, in both limiting cases of small and large wave numbers the low-frequency branch yields ion-acoustic oscillations with several different dispersion relations:

a) for long wavelengths ($k \to 0$)

$$\omega^2 \approx k^2 \left(\frac{Z\gamma_e T_e + \gamma_i T_i}{M}\right);$$

b) for short wavelengths ($k \to \infty$)

$$\omega^2 \approx k^2 \frac{T_i}{M}.$$

The second kind is damped rapidly, as will be shown below.

There is a difference between the long waves and short waves in that the electrons participate in the propagation of long waves; electrons move as though the electron mass were M/Z.

A very interesting result is obtained if we allow the **ion temperature** to approach zero for constant wave number and electron temperature in the ion dispersion relation. In this limiting case, for short waves the dispersion relation tends to

$$\omega^2 \approx \frac{mZ}{M}\omega_0^2.$$

Thus, the acoustic wave becomes the **ion plasma oscillation**. The frequency

$$\sqrt{\frac{mZ}{M}\omega_0^2} = \sqrt{\frac{n_e Z e^2}{\varepsilon_0 M}} = \sqrt{\frac{n_i Z e^2}{\varepsilon_0 M}}$$

is the **ion plasma frequency**. This is the frequency of electrostatic oscillations that arise as a result of ion displacement. The frequency depends on the charge and mass of the ions in the same way as the electron plasma frequency depends on the charge and mass of the electron.

In a real plasma with finite ion temperatures electrostatic ion oscillations are possible if the following of two conditions are satisfied:

$$\frac{k^2 T_e^2}{m} \gg \omega_0^2 \,; \qquad \frac{k^2 T_i^2}{M} \ll \frac{mZ}{M}\omega_0^2.$$

These conditions can be combined into the expression

$$T_i^2 \ll Z T_e^2.$$

Thus, if the ions are cold or have a high charge and if the electrons are hot, the ion branch is purely electrostatic over a broad range of wavelengths. In this range the ions oscillate with a constant electrostatic frequency. However, if we consider wavelengths outside this range (longer or shorter) the electrostatic oscillations become the ion-acoustic wave.

The frequency of electrostatic oscillations of cold ions is what would be expected if the ion displacement occurred in the presence of fixed electrons. Actually, the electrons are always far more mobile than the ions. However, if the electrons are hot, then their thermal motion "smears out" the electron density uniformly in space. In this case the ion oscillations are executed in a homogeneous electron background.

Plasma Oscillations in a Magnetic Field

A plasma in a magnetic field is capable of many different kinds of oscillations. The simplest of these are oscillations which propagate parallel or perpendicular to the magnetic field. In these two simple cases, in which the electric field is parallel to the magnetic field, the two modes of oscillation are uncoupled or independent. The waves that propagate parallel to the field are longitudinal electrostatic plasma oscillations; those that propagate perpendicular to the field are transverse electromagnetic waves. The magnetic field does not act on currents directed along it; therefore, it has no effect on oscillations of this kind.* Hence, it should be possible to probe a plasma with waves that propagate at right angles to the magnetic field but are polarized parallel to it.

This method of probing a plasma is used widely in plasma diagnostics. These waves can only penetrate the plasma if their frequency is higher than the plasma frequency (as in the absence of a magnetic field).

If the direction of propagation is parallel or perpendicular to the magnetic field, waves with arbitrarily directed electric fields can be decomposed into two waves: one, as described above, with its electric field polarized along the magnetic field, and a second with its polarization direction perpendicular to the field. Waves of the second kind, with electric field at right angles to the magnetic field, represent a new type of oscillation characteristic of plasmas located in a magnetic field.

If the frequency is small in comparison with the cyclotron frequencies (i.e., the ion cyclotron frequency, which is the lower) the plasma behaves as a conducting fluid and its actions are described by the equations of magnetohydrodynamics. In this frequency range the

---
* If we take thermal motion into account certain anomalies arise near the cyclotron frequency and its harmonics.

waves which propagate parallel to the magnetic field are called magnetohydrodynamic or Alfvén waves and those which propagate perpendicular to it are called magnetoacoustic waves. The physical nature of both of these types of oscillations can be clearly represented by using the frozen-in field concept. In both cases the magnetic field lines move together with the material. In the magnetoacoustic wave the plasma motion is in the direction of propagation. The mechanism of the wave is analogous to ordinary sound and consists of compression and expansion of the plasma together with the magnetic field frozen into it. The velocity of propagation can be found from the usual formulas for the sound speed if the magnetic pressure $B^2/2\mu_0$ is added to the ordinary kinetic pressure. The propagation speed of the magnetoacoustic wave is then given by

$$U_s^2 = \gamma \frac{p}{\rho} + \gamma_M \frac{B}{2\mu_0 \rho},$$

where $\rho = n_i M$ is the plasma mass density, and γ and γ_M are the adiabatic coefficients for the kinetic and magnetic pressures. The appropriate value for the magnetic adiabatic coefficient γ_M is 2. This can be inferred from both microscopic and macroscopic relations. From the microscopic point of view the magnetic field acts only on transverse particle motion, i.e., in the plane perpendicular to the field. The transverse motion has only two degrees of freedom. The adiabatic coefficient is given by

$$\gamma = \frac{f=2}{f},$$

where f is the number of degrees of freedom; hence, $\gamma_M = 2$. From the macroscopic point of view γ is determined by the relation

$$p \sim \rho^\gamma.$$

For a frozen-in field the density $\rho \sim B$ and the magnetic pressure $p_M \sim B^2 \sim \rho^2$. Thus, it follows that $\gamma_M = 2$. Substituting this value for γ_M we obtain the following relation for the sound speed perpendicular to the magnetic field:

$$U_s^2 = U_0^2 + \frac{B^2}{\mu_0 \rho} = U_0^2 + \frac{B^2}{\mu_0 n_i M},$$

where U_0 is the ordinary sound speed associated with the kinetic pressure

$$U_0^2 = \gamma \frac{p}{\rho} = \gamma \frac{T}{M}.$$

Now let us introduce the ratio of kinetic pressure to magnetic pressure:

$$\beta = \frac{2\mu_0 n T}{B^2}$$

where n is the total density of all the plasma particles. Then

$$U_s^2 = \frac{B^2}{\mu_0 \rho}\left(1 + \frac{\gamma}{2}\frac{n}{n_i}\beta\right).$$

For $\beta \to 0$ the speed of sound propagating at right angles to the magnetic field tends toward

$$U_A = \frac{B}{\sqrt{\mu_0 \rho}}.$$

This is the speed of pure magnetic sound in a cold plasma.

In the presence of a magnetic field the concept of a cold plasma acquires a completely defined meaning. We call a plasma "cold" if the kinetic pressure is much smaller than the magnetic pressure, i.e., $\beta \ll 1$. If this condition is satisfied, the kinetic pressure can be neglected.

The mechanism for oscillations which propagate along the magnetic field consists of the bending of the magnetic field lines together with the plasma which is "stuck" to them. The material motion here is perpendicular to the direction of wave propagation. These magnetohydrodynamic or Alfvén waves can be compared to the oscillations of a string. There is no analogy to the mechanics of an ordinary liquid or gas. Only in solid bodies is there an analogy with elastic oscillations. It might be said that the magnetic field imparts an elasticity to the plasma, thus making it like a solid in some sense.

It is remarkable that at low frequencies the propagation speed of the Alfvén wave along the field is just equal to the speed of magnetic sound in a cold plasma in spite of the fact that the physical mechanisms involved are completely different. This speed U_A is called the Alfvén speed.

In order to avoid misunderstanding we should call attention to a certain confusion in terminology. In the theories of elasticity and hydrodynamics, oscillations in which the material motions are parallel to the direction of propagation are termed longitudinal. Oscillations in which the vibrations are perpendicular to the direction of propagation are said to be transverse. In this mechanical sense the magnetoacoustic wave is longitudinal and the Alfvén wave is transverse. It appears then that the longitudinal oscillation propagates transverse to the magnetic field and the transverse wave propagates longitudinally along it. Note however that electromagnetic waves are considered transverse since their electric field lines are in the plane perpendicular to the direction of propagation. Electrostatic plasma oscillations are longitudinal since their electric fields are parallel to the propagation direction. In the electrodynamic sense Alfvén waves are transverse but magnetoacoustic waves are also transverse at low frequencies in the first approximation.

Dispersion near the Cyclotron Frequency

As long as the oscillation frequency is small in comparison with the cyclotron frequencies the propagation velocity does not depend on the frequency, i.e., there is no dispersion. In this range the phase and group velocities coincide. At frequencies comparable to the cyclotron frequencies, however, dispersion effects appear, i.e., the velocity of propagation starts to depend on the frequency. In this frequency range the phase velocity is not the same as the group velocity.

Especially strong dispersion is observed near the two characteristic frequencies at which the refractive index of the plasma tends to infinity, i.e., the phase velocity goes to zero. The analogous phenomenon in optics is called anomalous dispersion.

DISPERSION NEAR THE CYCLOTRON FREQUENCY

For waves propagating along the magnetic field the anomalous dispersion frequencies are the electron and ion cyclotron frequencies. Near these frequencies the electric field of waves with fixed propagation velocities do not have a fixed direction (plane polarized); instead, the direction of the electric field rotates. Such waves are called circularly polarized waves. The velocity of propagation of the waves depends on whether the rotation of the electric field is clockwise or counter clockwise. Media with such properties are called gyrotropic. Near the cyclotron frequencies the plasma is a gyrotropic medium.

A wave propagating along the magnetic field can be represented as a sum of two waves, both circularly polarized. One of these waves, in which the electric field rotates in the same sense as a positive ion in the magnetic field, is called the ordinary wave. It can propagate in the plasma at frequencies below the ion cyclotron frequency. At the ion cyclotron frequency the phase velocity of ordinary waves goes to zero; this frequency is the frequency of anomalous dispersion for ordinary waves. The second wave, in which the electric field rotates in the same direction as an electron in the magnetic field, is called the extraordinary wave. The electron cyclotron frequency is the frequency of anomalous dispersion for this wave.

Extraordinary waves can propagate in a plasma at all frequencies up to the electron cyclotron frequency. In the frequency range between the ion and electron cyclotron frequencies the propagation velocity of these waves passes through a maximum. This maximum value of the propagation velocity is equal to half the electron Alfvén speed:

$$U_{max} = \frac{1}{2} \frac{B}{\sqrt{n_e m \mu_0}}.$$

At frequencies much lower than the ion cyclotron frequency both the ordinary and extraordinary waves propagate with the same velocity, the Alfvén velocity. They can combine to form plane polarized waves in which the electric field points in a fixed direction. In this frequency range the gyrotropic properties of the plasma are not evident.

Waves propagating at right angles to the magnetic field also exhibit two frequencies at which anomalous dispersion occurs. These frequencies do not, in general, coincide with the cyclotron frequencies. These anomalous dispersion frequencies are called **hybrid** frequencies. Only in a very tenuous plasma, in which the plasma frequency is much lower than the electron cyclotron frequency, do the hybrid frequencies coincide with the cyclotron frequencies. At higher plasma densities, i.e., higher plasma frequencies, the hybrid frequencies are higher, being defined by:

a) the upper hybrid frequency

$$\omega^2 = \omega_0^2 + \omega_e^2;$$

b) the lower hybrid frequency

$$\omega^2 = \omega_i \omega_e \frac{\omega_0^2 + \omega_i \omega_e}{\omega_0^2 + \omega_e^2}.$$

We see from these equations that the upper hybrid frequency tends toward the plasma frequency in a dense plasma while the lower hybrid frequency approaches the geometric mean of the electron and ion cyclotron frequencies.

Waves can propagate in a direction exactly perpendicular to the magnetic field in the plasma at frequencies either lower than the lower hybrid frequency or higher than the upper hybrid frequency.* In the interval between these frequencies there is a region where propagation exactly perpendicular to the field is impossible. However, propagation can occur even at small angles to the perpendicular.

Waves that propagate across the field with frequencies below the lower hybrid frequency are called **magnetosonic waves**. At frequencies considerably below the ion cyclotron frequency the velocity of propagation approaches a constant value equal to the Alfvén speed. In this **magnetosonic region** the magnetosonic waves behave much like ordinary sound, relying on the magnetic pressure rather than kinetic pressure. The electric field and current in the magnetic sound region are perpendicular

---
* Here we assume that the plasma frequency is considerably higher than the electron cyclotron frequency.

both to the direction of the magnetic field and to the direction of propagation. As the lower hybrid frequency is approached the phase velocity of the magnetosonic wave approaches zero. The electric field and current of this wave also lie in the plane perpendicular to the magnetic field but have sizable components in the direction of propagation. From the electrodynamic point of view such oscillations are neither transverse nor longitudinal; they are elliptically polarized.

At frequencies higher than the upper hybrid frequency, high-frequency waves can propagate across the magnetic field. The electron pressure is important for these waves.

Oblique Waves and General Classification of Oscillations

We have looked at simple examples of waves propagating parallel and perpendicular to the field. In both cases the electric field can be oriented either in the plane perpendicular to the magnetic field, or parallel to the field. These two types of waves are independent of each other. In the general cases of oblique waves the electric field necessarily has components both parallel and perpendicular to the magnetic field. Thus these oscillations are much more complicated.

In a cold plasma in which the kinetic pressure can be neglected in comparison with the magnetic pressure, there are two kinds of waves. These waves are commonly referred to as the ordinary and extraordinary waves. If the direction of propagation is along the magnetic field the ordinary and extraordinary waves will be magnetohydrodynamic waves with variously directed circular polarization. In some texts, only waves with low frequencies in which there is no dispersion are referred to as magnetohydrodynamic or Alfvén waves; the ordinary and extraordinary waves are regarded as electromagnetic. However, we should keep in mind that these waves constitute a continuous spectrum, i.e., there is no division between them.

Extraordinary waves, propagating perpendicular to the field, are magnetic sound waves; the ordinary waves are electromagnetic with electric fields directed along the magnetic field.*

If we take the kinetic pressure into account, then the number of possible wave types increases and their properties become more complex. Magnetic sound waves become fast sound waves and the limiting nature of the lower hybrid frequency is removed. Plasma oscillations, which in a cold plasma have only the plasma frequency, become plasma waves, having a finite phase velocity which is a function of the frequency. Moreover, slow sound waves become possible; in the absence of a magnetic field these waves go over into ion-acoustic waves. Finally, the thermal pressure gives rise to the appearance of specific damping and sometimes to oscillation growth.

Propagation of Radio Waves through a Plasma

In the absence of a magnetic field electromagnetic waves can propagate in plasma only at frequencies higher than the plasma frequency. Waves at frequencies below the plasma frequency are reflected from the plasma boundaries. They can only penetrate a small distance of order c/ω_0. In the presence of a magnetic field waves polarized with their electric fields parallel to the constant external magnetic field behave in a similar manner. The magnetic field does not affect these waves.† The plasma particles move in the direction of the electric field and the magnetic field, which is also in this direction, exerts no influence on their motion.

We have seen that the plasma frequency is higher, the higher the electron density in the plasma. Therefore, only high-frequency waves can propagate in a dense plasma. For a given density there exists a cutoff frequency for the electromagnetic waves; this is

---

* We should mention that propagation exactly perpendicular to the field is, of course, a singular case. For small deviations from the perpendicular the particular peculiarities of this case vanish.
† See footnote on p. 113.

just the plasma frequency:

$$f_0 = 8.960 \sqrt{n}$$

where n is given in electrons/m^3.

Electromagnetic oscillations in which the frequency is lower (i.e., the wavelength is longer) than this limit cannot penetrate the plasma. On the other hand, for a given frequency there is an **electron density limit** in the plasma. This density is such that the given frequency is just equal to the plasma frequency. If the plasma density is lower than this limit the wave can propagate freely. The wave is reflected from plasmas with densities above the limit. On this basis a very valuable method has been devised for plasma diagnostics which is known as the **microwave probe** (Fig. 31). The plasma is exposed to a directed beam of electromagnetic waves. If the waves pass through the plasma and are detected by a receiver, then the plasma density must be lower than the limiting density. "Cutoff" of the electromagnetic signal indicates that the density is higher than this limit.

For the commonly used wavelength of 3 cm the limiting density is about 10^{18} electrons/m^3. If the plasma is located in a magnetic field waves with electric vector parallel to the magnetic field are used.

Wave propagation can be observed only at large distances from the transmitter, in the so-called wave zone. Thus, the

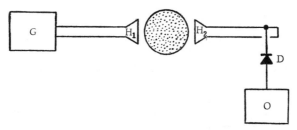

Fig. 31. Plasma probing by microwaves in the wave zone. The signal from a generator G goes through a waveguide to a horn H_1, aimed at the plasma. The signal transmitted through the plasma is received by a second horn H_2 and guided to a detector D, and is displayed on an oscilloscope O.

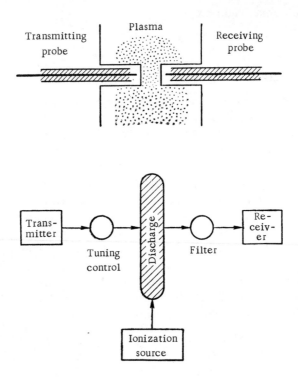

Fig. 32. Plasma probing by microwaves in the near zone.

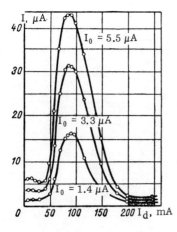

Fig. 33. The results of microwave probing in the near zone. The discharge current on which the plasma density depends is plotted on the horizontal axis. The intensity of the probe signal received is plotted on the vertical axis. Three curves are shown under identical conditions, but with different probe signal powers. The signals are detected at their maximum for a fixed plasma density at the point where the plasma frequency is equal to the probe signal frequency.

appearance of a limiting density and cutoff of the signal can occur only when the distance between the transmitting and receiving horns is greater than the wavelength. There are also other methods for high-frequency plasma probing in which this distance is small in comparison with the wavelength (Fig. 32). Transmitting and receiving probes are introduced into the plasma at a distance of 1-2 cm from each other. The working frequency is usually in the ultra shortwave range (meter waves).

At small distances from the transmitter (in the near zone) the maximum signal occurs where its frequency is equal to the plasma frequency (Fig. 33). This method is only suitable for very low densities (of the order of 10^{15} electron/m^3), such as are obtained in low-power gas discharges. At higher densities the plasma frequency corresponds to shorter wavelengths and it becomes difficult to locate the probes at a distance less than the wavelength.

The existence of a limiting density is very important for shortwave radio communication. Solar radiation creates a layer of ionized air (or plasma) in the earth's upper atmosphere. This layer is called the ionosphere. At night the ionization process is terminated and the plasma density diminishes because of recombination. In the daytime the density increases again under the action of solar radiation. The denser daytime plasma reflects shorter waves than the more rarefied night plasma. Thus, communication with these shorter waves becomes possible during daylight. At night these waves penetrate more deeply into the ionosphere and are absorbed.

Propagation of an electromagnetic wave in a plasma at a frequency higher than the plasma frequency usually means dispersion. The phase velocity of the wave U_{ph} is greater than the speed of light in vacuo; the group velocity U_g is smaller. We find

$$U_{ph} \cdot U_g = c^2.$$

This means that the index of refraction of the plasma $n < 1$. For frequencies much higher than the plasma frequency $n \to 1$, i.e., $U_{ph} \to c$. As the frequency approaches the plasma frequency $n \to 0$, i.e., $U_{ph} \to \infty$.

These relations are expressed mathematically by the disper-

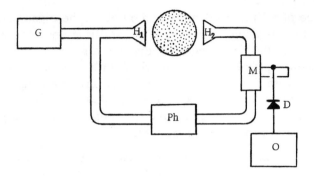

Fig. 34. Schematic diagram of a microwave interferometer. The signal from a generator G is split into two signals. One of these goes through the horn H_1, propagates through the plasma, and is received by a second horn H_2. The other signal is sent through a phase shifter θ. From the mixer M the sum of both the signals is received by a detector D the output of which is displayed on an oscilloscope O.

sion relation

$$\omega^2 = \omega_0^2 + k^2 c^2,$$

and the index of refraction $n = kc/\omega$ is

$$n^2 = 1 - \frac{\omega_0^2}{\omega^2}.$$

The phase velocity is

$$U_{ph} = \frac{\omega}{k}\sqrt{c^2 + \frac{\omega_0^2}{k^2}},$$

and the group velocity is given by

$$U_g = \frac{d\omega}{dk} = \frac{k}{\omega}c^2 = \frac{c^2}{U_{ph}}.$$

Another important plasma diagnostic technique is based on the dispersion of electromagnetic waves in the plasma. This is the microwave interferometer (Fig. 34). A guided radio signal with a frequency higher than the plasma frequency is divided into two beams. One beam is transmitted through the plasma; the other goes through a medium with index of refraction n = 1 (vacuum or air). Owing to the differences in phase velocity between these two beams there is a phase difference, which is measured by the interferometer. The actual phase velocity of the electromagnetic wave in the plasma is determined in this way. The dispersion relation can be used to find the density of the plasma.

Interferometer methods are more refined than probe methods. With one frequency the microwave method only gives the density at one point. The interferometer yields a continuous measurement of density if the plasma frequency is lower than the working frequency but not too much lower (for very high frequencies the phase velocity no longer depends on the density). If the plasma is in a magnetic field everything we have said applies only to waves on which the magnetic field has no effect. As noted above, these are waves in which the electrical field is parallel to the magnetic field. These waves can penetrate the plasma only if their frequencies are higher than the plasma frequency.

On the other hand, in the presence of a magnetic field low-frequency waves can penetrate the plasma; without the magnetic field these waves would be reflected from the plasma boundaries. In these waves the electric field must have components perpendicular to the magnetic field; otherwise the magnetic field has no effect on them.* If the plasma frequency is higher than the electron cyclotron frequency then only waves with frequencies lower than the electron cyclotron frequency can propagate through the plasma; there is an opaque range between the plasma frequency and the electron cyclotron frequency.

The important low-frequency wave that propagates in a magnetized plasma is a wave with circular polarization which propagates along the magnetic field. This wave propagates at all frequencies below the electron cyclotron frequency. At higher frequencies this wave is transformed into a magnetohydrodynamic (Alfvén) wave. Below the ion-cyclotron frequency Alfvén waves can have any direction of polarization, but in the region between the ion and electron cyclotron frequencies only waves with circular polarization can propagate.† In the case of propagation exactly

---
*See footnote on p. 113.
†See footnote on p. 118.

parallel to the field these waves are called extraordinary waves as noted above. The application of this term to oblique waves has not yet been generally agreed upon.

Waves with circular polarization that propagate in the ionosphere along with lines of the geomagnetic field are observed in the form of low-frequency radio noise. These radio signals are caused by lightning in the atmosphere; and are called atmospheric whistlers, or simply whistlers. These waves can propagate along the field lines for enormous distances from their place of excitation. Propagation requires that the electron cyclotron frequency in the earth's magnetic field must always be higher than the frequency of the wave.

Sometimes waves with circular polarization propagate along the magnetic field at frequencies lower than the electron cyclotron frequency and these are also called atmospheric whistlers.

If the plasma is in a strong magnetic field, the electron cyclotron frequency lies in the microwave range and waves with circular polarization can be observed at centimeter wavelengths. A recent method of microwave diagnostics is based on the observation of these waves in a plasma. Transmitting and receiving microwave horns are located along the axis of a magnetoplasma filament at right angles to the axis and to each other (Fig. 35). With this arrangement only signals with circular polarization are detected.

Plasma Resonators and Waveguides

If a plasma is enclosed in a chamber with highly conducting walls stationary oscillations (standing waves) can be established in it. To establish standing waves it is necessary to satisfy the condition of resonance: the propagation time of the wave must

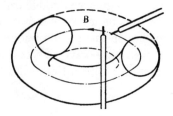

Fig. 35. Probing of plasma waves with circular polarization.

be an exact multiple of the oscillation period. The vessel (enclosed on all sides) in which the condition of resonance is satisfied in all directions is called a **resonator**. A long resonator with open ends is called a waveguide; these are traveling waves in the axial direction, but standing waves in the perpendicular direction. The resonance condition is only satisfied in the transverse direction in a waveguide.

Electrostatic plasma waves and ion-acoustic waves are unsuitable for resonators since they cannot be reflected from the walls.

In order that a wave be reflected from a highly conducting wall the electric field of the wave must be perpendicular to the direction of propagation, so that the wave must be transverse in the electrodynamic sense. The condition of resonance for such waves can be stated as follows: on the surface of the highly conducting wall the component of the variable electric field of the wave tangent to the wall must vanish. According to the law of electromagnetic induction it then follows that the total flux of the variable magnetic field of the wave through any cross section of the resonator must be equal to zero, i.e., the amplitude of the variable magnetic field must change sign inside the resonator.

Electromagnetic waves are used in high-frequency plasma resonators and waveguides. At frequencies higher than the plasma frequency the phase velocity of these waves approaches the speed of light. We can find the resonant frequencies first for the empty resonator, and then see how these change when the resonator is filled with plasma. This change of the resonant frequency or **resonator detuning** can be used to determine the density of the plasma. This is one of the most accurate methods of plasma diagnostics.

The possible operating regions of the resonator can be extended if the resonator or waveguide is placed in a magnetic field. The inherent symmetry of the magnetic field is cylindrical; thus, plasma resonators and waveguides for use in a magnetic field are usually made in the form of a cylinder. In addition to using high-frequency modes, it is also possible to use low-frequency magnetohydrodynamic and magnetoacoustic waves in resonators and waveguides in a magnetic field. If the resonance is limited to the micro-

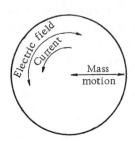

Fig. 36. Schematic diagram of oscillations in a magnetoacoustic plasma resonator.

wave range in the absence of a magnetic field, the magnetoacoustic resonance usually falls in the ultra short wave region. Magnetoacoustic resonances can be used to study the plasma, for plasma heating, generation of radio waves, and other applications in radio technology.

In Fig. 36 we show a schematic diagram of oscillations in a magnetoacoustic resonator at low frequencies in the magnetoacoustic range. The ac magnetic field is aligned with the axis of the cylinder, the mass velocity is radial, and the electric field and current are azimuthal. This configuration holds if the frequency is much lower than the ion cyclotron frequency. At higher frequencies, radial electric fields and currents appear. Oscillations with a purely radial propagation direction are possible only at frequencies above the lower hybrid frequencies. At still higher frequencies in an arbitrarily long cylinder we find that oblique waves can be excited which have their electric fields and currents along the axis of the cylinder.

The resonant frequencies of a cold plasma correspond to five different oscillation branches. Three of these are associated with high-frequency oscillations in which the basic role is played by electron motion. The other two oscillations are the low-frequency magnetosonic and magnetohydrodynamic oscillations. For each type of oscillation a volume with fixed dimensions and shape has an infinite number of resonant or natural frequencies.

The normal phase velocity of the oscillation becomes too large to satisfy the conditions of resonance for smaller volume dimensions or higher density. In this case the resonant frequencies approach the anomalous dispersion frequencies, near which the phase velocity diminishes rapidly. The character of the resonant phenomena for magnetosonic waves is determined by a dimensionless number called the **running electron number**

$$\Pi = \pi R^2 n = \frac{e^2}{4\pi\varepsilon_0 mc^2}.$$

This is the number of electrons along the length of a cylinder equal to the classical radius of the electron. It characterizes the ratio ω_0^2/k^2c^2. At a small value of the running electron number the resonant frequency of the magnetoacoustic oscillations approaches the lower hybrid frequency which, for these oscillations, is the anomalous dispersion frequency. At higher values the magnetoacoustic resonance is shifted toward lower frequencies into the magnetosonic range. In this region the resonant frequency is given by

$$\omega = \frac{\alpha}{R} \cdot \frac{B}{\sqrt{\mu_0 \rho}},$$

where α is a dimensionless number depending on the boundary conditions.

In the simplest case the quantity α appears as a root of the first order Bessel function. The amplitude of the magnetic field varies radially as a Bessel function of zero order and the amplitude of the electric field as a Bessel function of first order. The electric field goes to zero at the boundary of the cylinder. Thus, the variable magnetic field changes sign inside the cylinder in such a way that the total magnetic flux through any cross section of the cylinder is zero.

From electrodynamic considerations we see that anomalous dispersion must be accompanied by an absorption of energy, the mechanism for which we will examine below. The anomalous dispersion frequencies are often known as the resonant frequencies of the plasma. This resonant absorption is independent of the size and shape of the plasma volume; it also occurs in an unbounded plasma. Thus, it is necessary to distinguish resonance excitation as in the magnetosonic resonance just considered. Resonance excitation coincides with resonance absorption only when the running electron number is small.

If oscillations are excited in a plasma resonator by the application of an external voltage then resonant energy is fed into the plasma. In resonance absorption the plasma absorbs the energy. In the language of electrical engineering we can say that resonance absorption corresponds to a maximum active resistance in the plasma and resonance excitation, to a minimum reactance.

Excitation and Damping of Oscillations

All kinds of plasma oscillations can be supported if energy input is fed into a plasma from outside the system. If there is no such input oscillation in an equilibrium plasma must be damped by various energy-dissipation processes. However, real plasmas are very often not in complete thermodynamic equilibrium. In such nonequilibrium plasmas, oscillations can be excited as well as damped. Assume that a plasma is in hydrodynamic (but not thermodynamic) equilibrium and that oscillation are excited spontaneously. This means that the equilibrium state is unstable. Growing oscillations are merely another form of instability; we may call these oscillatory instabilities.* It follows from the general laws of thermodynamics that a condition of complete thermodynamic equilibrium cannot be unstable. Thus, an equilibrium state can be unstable to oscillation only in the hydrodynamic sense and not in the thermodynamic sense. Growing oscillations always appear as one of the links in the energy-transformation chain.

Oscillation damping can be caused either by collisions or by anomalous dissipation processes. Collisional damping is, in principle, very much the same in plasmas and ordinary gases. It occurs as a result of close interaction between particles. Anomalous dissipation is peculiar to plasmas and occurs as a result of the long-range interaction of particles with their self-excited fields.

Damping due to collisions takes place as a result of viscosity, Joule heat, and particle collisions. In a highly ionized plasma that supports oscillations in which electric currents are important, the dominant process is the electron–ion collision; this process gives rise to Joule heat.

In a fully ionized hot plasma, collisions are rare and the damping is due to anomalous dissipation. In the simplest kind of oscillation, electrostatic plasma waves, this process is known as Landau damping. It is based on phase resonance between the particles and the waves. Phase resonance can also result in growing oscillations.

---

* In the literature this form of instability is sometimes called (rather inappropriately) "overstability."

EXCITATION AND DAMPING OF OSCILLATIONS

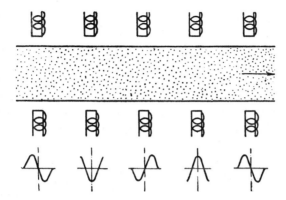

Fig. 37. An asynchronous plasma motor or dynamo. The curves under each coil illustrate the variation in phase.

The main features of phase resonance can be illustrated by the examples of an **asynchronous plasma motor** and **dynamo** (Fig. 37). Envision a tube containing plasma, surrounded by electromagnets. The current in these electromagnets are phase-shifted with respect to each other. The phase can be chosen so that the magnetic field moves along the tube with a fixed phase velocity. If the plasma is initially at rest, then the "freezing-in" of the moving field will cause the plasma to be accelerated. This will happen in the same way if the plasma is initially flowing along the tube with a velocity smaller than the phase velocity of the field. The field must be at rest in the plasma frame of reference, so that the plasma velocity will tend to match the phase velocity. The result is a plasma accelerator or motor in which electrical energy is transformed into kinetic energy of plasma motion.

Now let us see what happens if the initial velocity of the plasma stream in the tube is greater than the phase velocity of the field. In this case the moving field decelerates the plasma until the plasma velocity approaches the phase velocity of the field. As the plasma is braked, currents will be excited in it. These currents interact with the external magnetic field and give rise to a Lorentz force on the external windings. The mechanical energy liberated by the decelerating plasma is converted into electrical energy. Thus, we have a plasma generator (or dynamo).

The operation of the asynchronous plasma motor provides an analogy for anomalous damping; the asynchronous generator is analogous to oscillation growth in the plasma. Let a wave with fixed phase velocity be propagated in a plasma. Let us also assume that there are particles in the plasma which move with velocities such that the velocity component in the direction of wave propagation is equal to the phase velocity of the wave. These particles are then in phase resonance with the wave, i.e., the wave field acts on them with a phase which does not vary with time. Particles near phase resonance interact most intensely with the wave. If the particles are moving slightly slower than the wave, they acquire energy from it (as in the plasma motor); the energy lost in accelerating the particles causes the wave to decay. If the particles are moving slightly faster than the wave, they will release energy to the wave (as in the plasma generator) and the oscillation will increase in amplitude.

Analogous behavior is observed in the motion of discrete charged particles in a medium in which the speed of light is smaller than the speed of light in vacuo (this is the well-known Cerenkov effect). If a particle moves with a velocity greater than the speed of light in the medium, it absorbs light (Cerenkov absorption). Anomalous damping and growth of oscillations in a plasma can be regarded as an example of the Cerenkov effect.

Actually, there is a continuous distribution of particle velocities in a plasma. In this distribution there are some particles which are responsible for damping and some which are responsible for growth, i.e., there are particles with velocities which are slightly faster and slower than the phase velocity of the wave. If there are more slow particles than fast particles the oscillations are damped; if the reverse is true the wave will grow.

The mathematical distribution of particles with respect to their velocities is described by a distribution function $f(v)$. This function gives the fraction of the total number of particles having velocities near v. If the function $f(v)$ increases with v for $v = U_{ph}$, then the oscillations will grow; if the function decreases with v, the oscillations will be damped. The damping condition can be written

$$\left(\frac{\partial f}{\partial v}\right)_{v=U_{ph}} < 0$$

and the growth condition is

$$\left(\frac{\partial f}{\partial v}\right)_{v=U_{\text{ph}}} > 0.$$

In these inequalities v is not the total velocity, but its projection in the direction of U_{ph}. In thermodynamic equilibrium the distribution function always decreases monotonically with v, i.e., only damping is possible.

Near the anomalous dispersion frequency the phase velocity is very small. Even in a cold plasma there will be particles with velocities near the phase velocity. Therefore, anomalous dispersion is accompanied by anomalous damping even in a cold plasma. At frequencies far from the anomalous dispersion frequency the phase velocities of the waves are large and phenomena associated with phase resonance are possible only if fast particles are present. In anomalous damping the fast particles are present as a result of thermal motion due to plasma heating. Wave growth due to phase resonance occurs if there is a distribution of velocities associated with distinct groups of fast particles (there must be a "bump" on the distribution function as shown in Fig. 38). As an example, oscillations can be driven, or caused to grow, by the presence of an electron beam. The beam behaves as a separate group of fast particles which provides the necessary bump. A similar role can be played by electrons accelerated in the plasma by the action of electric fields. Such electrons are always present in a plasma when a current flows along the magnetic field. Since the deceleration of electrons due to collisions is less effective as the electron velocities increase, electrons moving in the forward direction can attain very high velocities. This **runaway** effect can cause growing instabilities in the plasma. This type of instability cannot be predicted by a continuum model. It is associated with the distribution of velocities over the particles as described by the equations of physical kinetics. This type of instability (in this case, the two-stream instability) is kinetic rather than hydromagnetic.

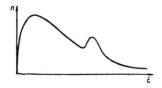

Fig. 38. A "bump" on the distribution function.

Instabilities can be excited by a variety of violations of thermodynamic equilibrium. In the case of kinetic instabilities such violations occur in the distribution of particles in velocity space. Excitation can also be caused by a nonequilibrium distribution in space, e.g., nonequilibrium temperatures. This question is still not completely settled, however.

Shock Waves in Plasmas

Supersonic flow in a nonconducting fluid can give rise to shock waves, i.e., surfaces at which the pressure, density, and stream velocity change abruptly over a short distance. This distance, i.e., the front thickness of the shock wave, is determined by dissipative processes such as viscosity and thermal conductivity. In a gas the thickness of the shock front is of the order of a mean free path.

In a plasma there can also be magnetohydrodynamic shock waves into which magnetic pressure enters together with the ordinary kinetic pressure. The role of the sound speed is played by the Alfvén speed and the role of the mean free path by the cyclotron radius or Debye length.

The structure of the shock front in a plasma is complicated and has an oscillatory character. In the absence of a magnetic field these oscillations are electrostatic, created by charge separation. In a magnetohydromagnetic shock wave the periodicity of the shock front is associated with plasma oscillations in the magnetic field. However, these are not linear oscillations (to which we have limited our discussions), but nonlinear oscillations with large amplitudes.

The existence of shock waves is possible only under conditions such that irreversible energy dissipation processes take place in the shock front. In a gas of neutral particles dissipation can only be due to collisions and therefore the thickness of the shock must be of the order of a mean free path. In a tenuous gas where the mean free path is long, shock waves are "spread out." In a plasma there are energy dissipation processes which do not require collisions. For this reason the shock thickness in a plasma can be much smaller than a mean free path and can be determined by other characteristic plasma lengths: the Debye shielding distance if there is no magnetic field, and the cyclotron radius if there is a magnetic field.

Dissipative processes not involving collisions make possible the propagation of collisionless shock waves in the rarefied plasma of interplanetary space. A solar flare generates a shock wave in the interplanetary plasma which, upon arrival at the earth, causes magnetic storms that start suddenly.

Random Processes

Up to this point we have only considered ordered motion, in which all of the plasma particles move in an ordered way, as soldiers in parade. This kind of motion is characteristic of a cold plasma. Thermal motion violates this ordered pattern. In a hot plasma the particles move without order, like school children. The thermal motion is chaotic. It is only subject to laws derived from the mathematical theory of probability.

It should be stressed that the magnetic field orders the particle motions and subjects them to rigid discipline. In the presence of a magnetic field the concept of a cold plasma is rigidly determined and has a real meaning. A plasma is called cold if its kinetic pressure $p = nT$ is small in comparison with the magnetic pressure $p_M = B^2/2\mu_0$. The ratio of these pressures is an important plasma parameter:

$$\beta = \frac{p}{p_M} = \frac{4\mu_0 nT}{B^2}.$$

At small values of β the plasma can be assumed to be cold; the role of the thermal pressure in such a plasma is negligibly small.

It is interesting to contrast the description of cold and magnetoplasmas. A plasma is considered a magnetoplasma if the cyclotron radius is small in comparison with the dimensions of the system*:

---

* Or the mean free path.

$$\frac{r_c}{R} = \frac{v}{\omega_c R} = \frac{v_i M}{ZeBR} \ll 1.$$

Here we consider the larger cyclotron radius, i.e., the ion cyclotron radius. Squaring the condition for a magnetoplasma we can write it in the form

$$\frac{M v_i^2 M}{Z^2 e^2 B^2 R^2} = \frac{T_{i\perp} M}{Z^2 e^2 B^2 R^2} \ll 1,$$

where R is the radius of the plasma (which we will take to be a cylinder extending along the field), and $T_i \approx M v_i^2$ is the temperature corresponding to transverse ion motion.

Multiplying the numerator and denominator by the ion density n_i, we can rewrite the magnetoplasma condition in the form

$$\frac{n_i T_{i\perp}}{B^2} \cdot \frac{M}{Z^2 e^2 n R^2} \ll 1$$

or

$$\beta \ll \frac{\mu_0}{4\pi} \frac{Z^2 e^2}{M} n\pi R^2.$$

Taking into account the relation

$$\epsilon_0 \mu_0 = \frac{1}{c^2}$$

we see that the first part of this inequality consists of a quantity called the running ion number. This is the total number of ions over the length of the cylinder equal to the classical radius of the ion $Z^2 e^2/4\pi\epsilon_0 Mc^2$. If the total number of ions is large the cold plasma will clearly be a magnetoplasma.

The most important effect of kinetic (or thermal) motion lies in the area of **transport phenomena**: transport of heat, matter, and momentum. The transport of matter by thermal motion is is called **diffusion**; heat transport is called **thermal conduction**, and momentum transport is called **viscosity**. In a plasma with finite thermal pressure there can also exist charge transport, i.e., electrical currents, since the electrical conductivity of the plasma can be related to a number of the other transport coefficients. However, the electrical conductivity does not have as simple an interpretation as the coefficients of diffusion, thermal conductivity, and viscosity.

The basic law for all random processes is the dispersion law. Let us examine the random quantities that characterize, for example, the position or velocity of a particle which contributes to the random thermal motion. We denote this quantity by f. It can vary in a completely random manner. For simplicity we shall first let f change discontinuously and assume that in each "jump" it changes by an amount Δf, where Δf can be positive or negative with equal probability. Summing over all these jumps we have

$$\Sigma \Delta f = 0.$$

Thus, the average value of f does not change. However, the squares of all variations are positive, and the sum of these squares

$$\Sigma (\Delta f)^2 > 0$$

increases proportionally with time. This sum is called the dispersion. It serves as a measure of the range over which the quantity f is "smeared out" about its constant average value. Since the dispersion increases directly with time we can write

$$\Sigma (\Delta f)^2 = Dt.$$

The coefficient of proportionality D is called the **diffusion coefficient**.

The dispersion law can be generalized to cover the case in which the quantity f is not only determined in magnitude, but also in direction, i.e., vector phenomena. In this case the dispersion law becomes the **law of random walk**. This meandering of particles in ordinary space leads to transport processes. In velocity space it causes an energy exchange between different degrees of freedom and leads to the establishment of thermal equilibrium.

The Drunkard's Walk

The random walk in ordinary space can be illustrated by the following graphic example (Fig. 39). Imagine a drunkard walking through a wood with no particular purpose in mind. Each time he encounters a tree he turns in a completely arbitrary direction. His path covers an area which increases in proportion to the time. Assume that the area can be taken as a circle whose center remains stationary and whose radius increases with t.

The area occupied by the meanderings (the migration area) can be written as

$$S = Dt.$$

The constant of proportionality D is the diffusion coefficient. If each time the drunk changes his direction he has covered a straight path of length l since his last encounter, then

$$S = \Sigma l^2.$$

Thus, the diffusion coefficient is

$$D \approx \frac{l^2}{\tau},$$

where τ is the time between two successive collisions. The length l in the problem of the drunkard's walk is simply the distance between neighboring trees. In the more general random-walk problem this is the displacement of particles after each collision or, mean

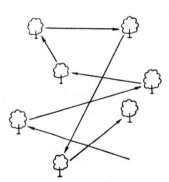

Fig. 39. The drunkard's walk.

free path. In the problem of the drunkard's walk all of the walking takes place in a plane. However, for walk in a volume and also the case in which the motion is limited to a line, the dispersion law is always effective, i.e., the square of the linear dimension of the walk region increases in proportion to the time

$$L^2 \approx Dt.$$

Whence follows the basic relation for all diffusion processes: the penetration depth for a time t is

$$L \approx \sqrt{Dt}$$

and the time for penetration to a given depth L is

$$t \approx \frac{L^2}{D}.$$

The quantity L^2 has the dimensions of an area and it is called the migration area even when this area does not have a geometrical interpretation. The constant of proportionality D is always called the diffusion coefficient and, in order-of-magnitude terms, is equal to the product of the path length and the velocity.

The Mean Free Path and the Collision Cross Section

An uncharged particle can only have its direction of motion changed by a collision. Between collisions the particle moves in a straight line with constant velocity v over a distance given by

$$l = v\tau,$$

where τ is the time between collisions. The diffusion coefficient can then be written

$$D = lv.$$

If colliding particles behave as a rigid body of transverse cross section Q, then a particle moving with velocity v in a medium with n particles per unit volume experiences nvQ collisions per unit time. The time between successive collisions is just the reciprocal of this quantity

$$\tau = \frac{1}{nvQ},$$

and the path length is

$$l = \frac{1}{nQ}.$$

The denser the medium, the shorter the path length.

If we consider a plasma with particles of different types, then the number of collisions is additive, i.e., the inverse collision times or path lengths must be added. For a partially ionized plasma we have

$$\frac{1}{\tau} = \frac{1}{\tau_a} + \frac{1}{\tau_i} = v(n_a Q_a + n_i Q_i),$$

where n_a and Q_a are density and cross section for neutral particles and n and Q are the corresponding parameters for ions. The electrical conductivity of a partially ionized plasma is

$$\sigma = \frac{ne^2}{m}\tau = \frac{\frac{ne^2}{m}}{\frac{1}{\tau_a} + \frac{1}{\tau_i}},$$

where

$$\frac{1}{\sigma} = \frac{1}{\sigma_a} + \frac{1}{\sigma_i},$$

and

$$\sigma_a = \frac{ne^2}{m}\tau_a = \frac{ne^2}{mvn_a Q_a},$$

$$\sigma_i = \frac{ne^2}{m}\tau_i = \frac{ne^2}{mvn_i Q_i}.$$

Collisions with Neutral Particles

Neutral particles can, to a reasonable approximation, be assumed to have a certain geometrical cross section. The particles do not appear to be rigid however, i.e., the cross section always depends on the velocity. For neutral particles this dependence is determined solely by invariant properties of the particles and cannot be stated in a general way. Collisions can be either elastic or inelastic. In elastic collisions only an exchange of kinetic energies takes place; in inelastic collisions the kinetic energy of the colliding particles is partially or wholly changed into another form of energy. This energy transformation process can include excitation, ionization, or charge exchange. In excitation the kinetic energy is spent in shifting electrons in one or more of the atoms or molecules participating in the collision into higher quantum energy levels. This energy is subsequently released by radiation the the form of light. If the available kinetic energy exceeds the energy of the highest quantum level ionization becomes possible. Finally, if an atom collides with an ion of its own species, there is a high probability that charge exchanges will take place. This is a very important process in plasma physics. The ion extracts an electron from the atom, and the ion becomes an ion (Fig. 40). It would appear that nothing has changed. The process is reminiscent of the story, told by Koz'moi Prutkov, about the soldiers, Schmidt and Schultz, who were late in returning from leave and decided to confuse their commander. Schmidt said he was Schultz and Schultz said he was Schmidt. In a plasma "name exchange" (i.e., charge exchange) is very important.

An ion can be accelerated by electric fields and confined by magnetic fields. If a fast ion acquires an electron from an atom it becomes a fast atom, on which the magnetic field has no effect. An atom cannot be confined in a magnetic trap. It will hit the wall and lose the kinetic energy which it acquired from the electric field. The energy of the ion produced in the charge exchange is small; it must be reaccelerated. Thus, the process of charge exchange is one of the chief sources of trouble in the effort to obtain a hot plasma.

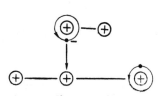

Fig. 40. Charge Exchange.

The processes of excitation and ionization have a threshold. For these processes the collision energy must be equal to or greater than some given threshold value. The charge exchange process is most probable when it is resonant, i.e., when the ion obtains an electron from an atom of the same type. Such resonant charge exchange occurs with high probability at low energies.

The probability of the various collisional processes is commonly characterized by the corresponding effective cross section: the cross section for excitation Q_e, the cross section for ionization Q_i, and the cross section for charge exchange Q_c.

The cross section for a given process is defined in such a way that a particle, moving with velocity v in a medium of density n experiences vQn events per unit time. The sum of these three cross sections is the total cross section for inelastic processes:

$$Q = Q_e + Q_i + Q_c.$$

The cross sections for excitation and ionization are zero below their energy thresholds. Above the threshold they increase rapidly with energy, pass through a maximum, and then fall off fairly quickly. The cross section for charge exchange decreases more gradually with increasing energy. The maximum value of the ionization cross section is approximately πa^2, where a is the radius of the atomic electron orbits. The cross section for resonant charge exchange is much greater than the geometrical dimensions of the atom since the ion need not necessarily come in close contact with the atom in order to remove an electron (see Fig. 40).

Experimentalists like to use the so-called collision "probability" P instead of the cross section Q. This is the number of collisions of a given kind in a 1 cm path at 1 torr initial pressure (at normal temperature). The probability and the cross section are related by

$$P = \frac{L}{760} Q = 3.5357 \cdot 10^{16} \cdot Q,$$

where L is the Loschmidt number, i.e., the number of particles in 1 cm^3 of gas at standard temperature and pressure (STP). Figure 41 shows the ionization probability and Fig. 42, the resonant

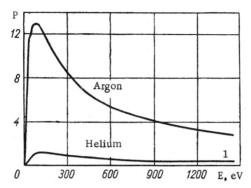

Fig. 41. The probability of ionization for helium and argon as a function of the electron energy.

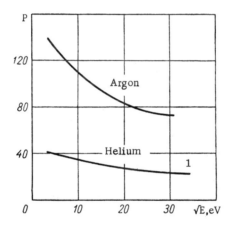

Fig. 42. The probability of charge exchange for helium and argon as a function of the collision energy.

charge exchange probability, for helium and argon as functions of the energy of the colliding particles. Since the probability of any process is proportional to the cross section, the curves for the cross sections are of the same form and only the vertical scale is changed. The cross section for the ionization of argon is larger than for helium because of the larger radius of the outer electron orbits. The cross section for resonant charge exchange is about ten times larger than the ionization cross section. If we convert from probability to cross section, we find that the ionization cross section is about 10^{-20} m^2 and the charge exchange cross section is about 10^{-19} m². Probabilities are simpler to use since they are expressed in more convenient numbers (one to a few hundred).

Elastic collisions lead to particle scattering and energy exchange and are important in the electrical conductivity and transport processes. The cross section for elastic collisions is usually of the order of the geometrical dimensions of the atoms or molecules involved. Anomalously small cross sections are found for elastic scattering of slow electrons by inert gas atoms: argon, krypton, and xenon. Here, thanks to the quantum mechanical phenomenon of electron defraction, the electrons are hardly scattered by the atoms. They move through the gas almost as though it were a vacuum. This phenomenon is called the Ramsauer effect. It is observed at electron energies considerable below the threshold for all inelastic processes.

Coulomb Collisions

Charged particles interact according to the Coulomb force law, according to which the interaction strength diminishes slowly with distance. Charged particles, moving in a plasma, constantly interact with other charged particles via this long-range force. For this reason the path of a particle is not a broken one (with discontinuous derivatives) as is the case with neutral particles, but a smooth curve (Fig. 43). However, a detailed analysis shows that thermal motion and transport processes can still be described in terms of a collision theory by introducing the concept of Coulomb collisions. This means that the actual curved path of the particles can be replaced by a series of broken lines (the dashed line in Fig. 43). Each time the direction of a particle changes by approximately 90° as a result of continuous interactions we say that the particle has experienced a Coulomb collision; this corresponds to a change in direction in the dashed line. In order to describe thermal motion in a plasma quantitatively by a

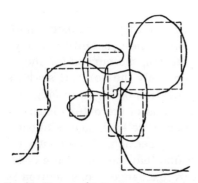

Fig. 43. Diagram for Coulomb collisions.

collision theory, we must determine an effective cross section for Coulomb collisions. We will call this the Coulomb cross section. It consists of two factors; these take account of the near and remote interactions.

Near interactions are illustrated by cases where a single interaction between two particles leads to a sharp deflection. In order for this to occur it is necessary that the potential energy of the Coulomb interaction be of the same order as the kinetic energy of the colliding particles:

$$\frac{Z_1 Z_2 e^2}{4\pi\varepsilon_0 r} = Mv^2$$

(the coefficient 1/2 is omitted from the right-hand term as it is unimportant for this estimate). Thus, the distance for a near interaction is

$$b \approx \frac{Z_1 Z_2 e^2}{4\pi\varepsilon_0 Mv^2}.$$

This distance is known as the impact parameter. It must be noted that only the kinetic energy of relative particle motion plays a role in the collision; thus, v is the relative velocity and M is the reduced mass

$$M = \frac{m_1 m_2}{m_1 + m_2}.$$

For interactions between electrons and ions the reduced mass is approximately equal to the electron mass; the impact parameter is then

$$b \approx \frac{Ze^2}{4\pi\varepsilon_0 mv^2},$$

where m is the electron mass and Z is the charge number of the ion.

The effective cross section for a near or binary interaction is the area of a circle of radius b, i.e., πb^2.

However, in addition to sharp deflections, the direction of motion of the particles can also be changed by remote interactions which produce a gradual curvature in the path. Theory shows that the complete Coulomb cross section is obtained by multiplying the cross section for close interactions by the **Coulomb logarithm**:

$$\tilde{Q} = \pi b^2 \ln \Lambda.$$

The Coulomb logarithm is of order 10, under conditions common to plasma; therefore, we see that remote interactions are considerably more important than close interactions.

The impact parameter for near collisions b is inversely proportional to the kinetic energy of the particles. In a thermal plasma it is inversely proportional to the temperature. Consequently, the cross section for close interaction decreases with increasing temperature, going as $Q \propto 1/T^2$. The Coulomb logarithm (like all logarithms) is only weakly dependent on the velocity or energy of the particles. Hence, the basic property of the Coulomb cross section is an abrupt decrease with increasing particle velocity. In a thermal plasma the Coulomb cross section is inversely proportional to the square of the temperature. If, under the influence of an electric field, a group of electrons are separated from their normal distribution and acquire high velocities, for these electrons the Coulomb cross section becomes vanishingly small as the velocity increases further and the cross section diminishes progressively. Such "runaway" electrons can, in the final analysis, be accelerated in a plasma as though it were a vacuum. It is necessary that the electric field be directed along any magnetic field that may be present, i.e., the current must be parallel (or antiparallel) to the field lines.

The quantity Λ, the Coulomb logarithm, is the ratio of the maximum and minimum impact parameters:

$$\Lambda = \frac{r_{max}}{r_{min}}.$$

The maximum impact parameter is of the order of the Debye shielding distance

$$r_{max} \approx h.$$

For distances larger than this the Coulomb potential decays to zero rapidly and the interaction vanishes.

The minimum impact parameter can, depending on the relative velocity of the particles, be either classical or quantum mechanical. At low velocities, quantum phenomena are not important (we call this case **quasi-classical**) and the radius for a near interaction can be taken for the minimum impact parameter

$$r_{min} \approx b.$$

For electrons

$$r_{min} \approx \frac{Ze^2}{4\pi\varepsilon_0 mv^2}.$$

However, this distance must be greater than the deBroglie wavelength:

$$r_{min} \geqslant \frac{\hbar}{mv}.$$

If the opposite is true, quantum-mechanical effects enter and it is necessary to take the deBroglie wavelength as the minimum impact parameter

$$r_{min} \approx \frac{\hbar}{mv}.$$

Quantum effects can only make a significant contribution for electrons. Thus, the minimum impact parameter for electrons will be the larger of two quantities: the close interaction impact parameter b or the electron wavelength $\hbar/mv$. The minimum impact parameter is purely quantum mechanical if

$$\frac{Ze^2}{4\pi\varepsilon_0 mv^2} \ll \frac{\hbar}{mv},$$

or

$$\frac{Ze^2}{4\pi\varepsilon_0 \hbar v} \ll 1.$$

This condition is identical with the condition of applicability of the Born approximation in quantum mechanics. The quantum or Born case therefore obtains for high electron velocities. In a thermal plasma the temperature must be well above 400,000°K (~40 eV) in order for quantum effects to be important.

The **classical radius** of a particle is the near impact parameter for a kinetic energy equal to the rest energy. In accordance with the theory of relativity, the rest energy is the mass multiplied by the square of the speed of light: $E_0 = Mc^2$. Thus, the classical radius is

$$r_0 = \frac{Z^2 e^2}{4\pi\varepsilon_0 M c^2} \ .$$

The minimum impact parameter can then be written

$$b = r_0 \frac{E_0}{E}$$

and the Coulomb cross section becomes

$$\tilde{Q} = \pi r_0^2 \left(\frac{E_0}{E}\right)^2 \ln \Lambda.$$

For electrons

$$r_0 = \frac{e^2}{4\pi\varepsilon_0 m c^2}.$$

The cross section for scattering of light by particles (Thomson scattering) is

$$Q_0 = \frac{8}{3} \pi r_0^2.$$

The Coulomb cross section can be expressed in these terms as

$$\tilde{Q} = \frac{3}{8} Q_0 \left(\frac{E_0}{E}\right)^2 \ln \Lambda.$$

Here E is the kinetic energy of the particles.

The Establishment of Thermal Equilibrium

In a tenuous plasma the velocities and energies of the particles can sustain a nonequilibrium distribution for a considerable period of time. Any difference between the average energies of the electrons and the ions can exist for an especially long time, so that it is often necessary to describe a plasma by two temperatures,

THERMAL EQUILIBRIUM

the ion temperature T_i and the electron temperature T_e. The establishment of thermal equilibrium at each point in space occurs as a result of random velocity fluctuations. The time required to establish equilibrium is known as the relaxation time. It is necessary to distinguish between three different relaxation times. The establishment of an equilibrium in the electron velocity distribution occurs first, i.e., this process has the shortest relaxation time τ_e. A much longer time is required for the ion velocities to reach an equilibrium distribution. We shall denote this time by τ_i. An even longer time is required for complete energy exchange between the electrons and ions so that the two temperatures become equal. The time required is τ_{ei}. For identical final temperatures these times are related in the following way:

$$\frac{\tau_e}{\tau_i} = Z^4 \sqrt{\frac{m}{M}}; \quad \frac{\tau_i}{\tau_{ei}} = \frac{1}{Z^2}\sqrt{\frac{m}{M}}.$$

If the plasma is initially not in equilibrium, the concept of temperature has no meaning for time τ_e. If we consider a time greater than τ_e but smaller than τ_i we can speak of an electron temperature but not an ion temperature. If we then consider a period of time longer than τ_i but smaller than τ_{ei} two different temperatures, the electron temperature and the ion temperature, both have meaning; after a time τ_{ei} the plasma can be described by a single temperature.

The magnitude of the relaxation time can be estimated from the equation for the collision time:

$$\tau = \frac{1}{nvQ}.$$

If we substitute the Coulomb cross section for Q and replace v by the root-mean-square velocity of the thermal motion,

$$v = \sqrt{3\frac{T}{M}},$$

where T is the temperature in energy units. The magnitude of the velocity v is taken from the equilibrium conditions since this condition is approached by the relaxation process. An exact calculation gives a value for the relaxation time which is eight times smaller than our estimate:

$$\tau = \frac{1}{8\pi n v r_0^2 \ln \Lambda}.$$

Thus,

$$\tau_e = \frac{\sqrt{m(3T_e)^3}}{8\pi n e^4 \ln \Lambda},$$

$$\tau_i = \frac{\sqrt{M(3T_i)^3}}{8\pi n Z^4 e^4 \ln \Lambda}.$$

For collisions between electrons and ions only a small portion of the energy is transferred from one to the other because of the large mass difference; this energy fraction is of the order of the mass ratio m/M. Therefore, the time required to establish thermal equilibrium between the electrons and the ions can be written

$$\tau_{ei} = \frac{M}{m} \frac{\sqrt{m(3T)^3}}{8\pi n Z^2 e^4 \ln \Lambda}.$$

If there are only singly charged ions in the plasma, then τ_i is $\sqrt{M/m}$ times larger than τ_e and τ_{ei} is much larger than τ_i.

Transport Processes in a Magnetic Field

In a magnetic field all transport processes are anisotropic, i.e., they behave differently in different directions. Particles move freely along the magnetic field as though it were not there. In this direction, after each collision particles move a distance

$$l = v\tau$$

and the diffusion coefficient is

$$D_\parallel \approx \frac{l^2}{\tau} \approx lv.$$

In the limiting case of a strong magnetic field the particles do not move freely across the field; instead they gyrate in cyclotron circles. As a result particle collisions can only cause the particle to jump from one circle to another. The displacement of a particle after each collision will then be of the order of the cyclotron radius

TRANSPORT PROCESSES IN A MAGNETIC FIELD

$$r_i = \frac{v}{\omega_c}$$

and the diffusion coefficient perpendicular to the field will be

$$D_\perp = \frac{r_c^2}{\tau} \approx \frac{v^2}{\omega_c^2 \tau}.$$

The ratio of the transverse and longitudinal diffusion coefficients in a strong magnetic field is

$$\frac{D_\perp}{D_\parallel} \approx \frac{v}{\omega_c^2 l \tau} = \frac{1}{\omega_c^2 \tau^2}.$$

Let us look at two limiting cases:

a) in the absence of the field

$$D_\perp = D_\parallel,$$

b) in a very strong field

$$D_\perp = \frac{D_\parallel}{\omega^2 \tau^2}.$$

Both of these cases can be included in a single equation:

$$D_\perp = \frac{D_\parallel}{1 + \omega_c^2 \tau^2}.$$

A plasma, in which the period of the cyclotron gyration is much shorter than the time between collisions, is called a **magnetoplasma**. The criterion for this condition is, as we have seen,

$$\omega_c \tau \gg 1.$$

In a magnetoplasma the anisotropy of the transport processes is highly pronounced. The transverse diffusion coefficient, i.e., the coefficient for diffusion at right angles to the field, is inversely proportional to the square of the cyclotron frequency, i.e., the square of the magnetic flux density. The same is true of the other transport coefficients, in particular the thermal conductivity and the electrical conductivity.

Actually, transport processes in a magnetoplasma are not as anisotropic as indicated above because of a number of complicating circumstances. Electrical currents in a plasma do not arise only as a result of electric fields, but also from inertial and pressure forces. These forces often remove the anisotropy of the electrical conductivity. Diffusion processes are also complicated by anomalous diffusion caused by plasma instabilities. Thus, as we see, the difficulties encountered in the attempt to confine a plasma in a magnetic trap are of a very complex nature.

Ambipolar Diffusion

In an electrically isolated plasma the ions and electrons cannot diffuse independently of each other since this would violate neutrality. Any small deviation from neutrality is evidenced by the presence of an electric field. This field acts to inhibit any further charge separation. As a result of the "lagging" particles, the motion of those particles that move ahead is retarded. The situation is similar to that which occurs on guided tours, in which the entire group must observe the pace of the slowest member. Of course, the trailing particles also have a tendency to be pulled to the front, but to a lesser extent than the tendency for leading particles to be held back. In the simplest case, if the diffusion coefficient for particles of one sign is very much larger than for particles of the opposite sign, the total diffusion coefficient will only be half as large as the largest individual coefficient. This combined diffusion of oppositely charged particles is called **ambipolar diffusion**.

Without a magnetic field (or along a field) the electrons diffuse much more rapidly than the ions since they are light and much more mobile. In this case the coefficient for ambipolar diffusion is twice the diffusion coefficient for the ions. The inverse relationship is obtained for diffusion across a strong magnetic field. In this case the diffusion coefficient for the ions is much larger (because of the larger cyclotron radius) and the coefficient for ambipolar diffusion is twice the diffusion coefficient of the electrons.

Under certain conditions diffusion perpendicular to the magnetic field is not ambipolar. This effect arises when the charge

AMBIPOLAR DIFFUSION

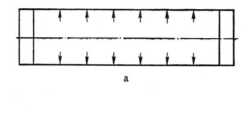

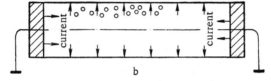

Fig. 44. Ambipolar (a) and Simon (b) diffusion.

separation is "short circuited" by currents that flow along the magnetic field. As an example we consider a plasma in a cylindrical tube with an axial magnetic field directed along the tube (Fig. 44). The plasma particles diffuse across the magnetic field into the wall. It can be shown that the diffusion rate depends strongly on the material used to block off the ends of the tube. If insulated ends are used the charges that arise in the plasma are not neutralized. In this this case the diffusion is ambipolar. The diffusion rate will be twice the rate of electron diffusion and much less than would be the case for ions diffusing across the field alone. However, if the insulated ends are replaced by metal ends the character of the diffusion is completely changed.* Now the ions can diffuse at their characteristically high rate and the electrons bring up the rear. Excess electrons can drift freely along the magnetic field to the metal end plates and through them to ground. This means that a charge compensating current flows along the magnetic field. In this case the diffusion will cease to be ambipolar and its rate will then be determined by the larger of the two particle diffusion coefficients (Simon diffusion).

This example is instructive. Evidently the diffusion coefficient, which is assumed to be a physical constant for other states of matter, depends on the external conditions in a plasma. This sort of thing occurs frequently for physical quantities in a plasma, e.g.,

---

* This effect is stronger if the walls of the cylinder are also made of a conducting material.

the electrical conductivity, etc. It is risky to accept a physical constant without question since the behavior of matter in the plasma state is more complex than in other physical states.

A Recent Plasma Experiment

It would be useful to conclude this book by making a survey of applications. However, the time is not yet right for this. Many interesting technical possibilities have already appeared in the course of our discussion of various physical problems. The most important technical applications of plasma physics are yet to emerge from the experimental or developmental stage in the laboratory. Thus, in place of a survey of applications, we will close this book by describing a typical plasma experiment.

In the classical physics of gaseous discharges a plasma was produced by a direct current between metal electrodes. The degree of ionization was very low. The magnetic field employed, if any, was very weak. The most important effects which appeared in the discharge were physicochemical in nature and were involved with the electrode material. These phenomena are more properly in the realm of physical electronics than of plasma physics.

On the other hand, in recent experiments the tendency has been to obtain a dense plasma in an electrodeless inductive discharge and to apply powerful magnetic fields in order to exploit the properties of frozen fields and magnetic pressure. To obtain such high fields and energy densities we use pulse processes that last for thousandths or millionths of a second (milliseconds or microseconds). To obtain such a pulse the electrical energy is stored in a capacitor bank. The recording of processes occurring in the plasma is accomplished by oscillographs and high-speed photographic techniques. In order to obtain a high degree of ionization is is necessary to work with low-density gases; the initial pressures are usually a few hundredths or thousandths of a torr.

The experiment (Fig. 45) is concerned with the compression of a plasma column by a rapidly rising longitudinal magnetic field (θ-pinch). The capacitor bank is charged through a rectifier. The

A RECENT PLASMA EXPERIMENT

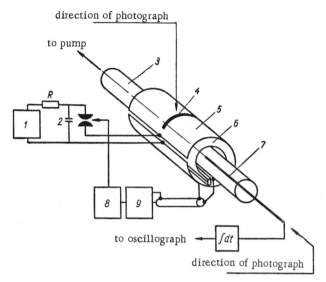

Fig. 45. Schematic diagram of the experimental arrangement.
1) Rectifier; 2) capacitor bank; 3) quartz tube; 4) slit in coil;
5) long coil; 6) auxiliary winding; 7) magnetic probe; 8) pre-ionizer generator; 9) trigger circuit.

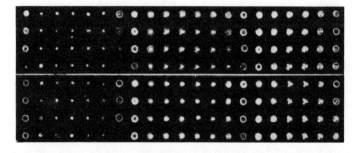

Fig. 46. Framing photographs of the θ-pinch taken end-on.

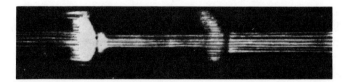

Fig. 47. A streak photograph of the θ-pinch taken side-on.
Time runs from left to right.

gas, contained in a quartz tube, is preionized by a high-frequency discharge from a low-power generator through an auxiliary winding. At a small fixed time after the preionizing discharge the capacitor bank is triggered, discharging a large current pulse through a long single-turn coil. This discharge establishes a rapidly rising axial magnetic field. The magnetic field inside the plasma is measured by magnetic probes. Oscillograms of this field have been shown in Fig. 30. High-speed photographs are taken end-on and through a slit in the coil. Framing photographs are taken through the end of the tube. These frames are equispaced in time to give a time history of the behavior as viewed end-on. A series of typical frames is shown in Fig. 46. From the side, streak photographs are taken on film by a rotating drum camera (Fig. 47). A comparison of the photographs with the magnetic field oscillograms yields information on the behavior of the plasma column.

The capacitor discharge is oscillatory in nature (see Fig. 30b). In the first quarter-cycle the growing external field compresses the plasma column together with the field of the same sign, which is frozen into it. In the second half-cycle the external field changes sign and the process of opposed field diffusion begins. A neutral layer is formed, as was discussed earlier. The first part of the oscillograms shown in Fig. 30 applies to this half-cycle. The shock wave which arises as the neutral layer is compressed drives the plasma column into oscillation and causes free magnetosonic oscillations to occur. Compression of a neutral layer by opposed fields in experiments of this type (with very fast compression) have yielded the highest temperatures ever obtained in the laboratory. By rough estimate these temperatures are of the order of tens of millions of degrees. Subsequent rapid acceleration of the plasma leads to instability of the plasma column. Along the magnetic field lines longitudinal striations appear; these are seen in both the end-on and side-on pictures. The "life" of the plasma is usually terminated by these instabilities. This "disease" is as bad for a plasma as cancer or heart disease is for humans. The struggle to overcome instabilities is a most important part of the continuing work of plasma physics.

Index

Absorption coefficient, 38
Acceleration, plasma, 55
Adiabatic invariant, 74
Adiabatic motion, 72
Adiabatic trap, 76
Alfvèn speed, 116
Alfvèn wave, 114
Ambipolar diffusion, 152
Anisotropic pressure, 29
Anomalous dispersion, 116
Anomalous dissipation, 130
Arc, 25

Beta (ratio of gas-kinetic pressure to magnetic pressure), 48
Bound—bound radiation, 36
Bremsstrahlung, 37

Cathode fall, 25
Centrifugal drift, 79
Cerenkov effect, 132
Charge exchange, 35, 141
Chromospheric flare, 104
Conductivity,
    longitudinal, 64
    transverse, 64
Conducting-fluid model, 45, 49
Contact potential, 23
Convection, 92
Coloumb collision, 144

Cross section, 142
Cusp, 100
Cyclotron frequency, 66
Cyclotron gyration, 39

Debye length, 32
Degeneracy, 33
Diagnostics, 11
Diamagnetic current, 86
Diffusion coefficient, 137
Diffusion time, magnetic field, 49
Dispersion, 106
Dispersion relation, 109
Drift, 46, 68

Electric drift, 71
Electrodeless discharge, 24
Electromagnetic pump, 55
Electron avalanche, 9
Electrostatic energy, plasma, 31
Equation of state, plasma, 27
Extraordinary wave, 117

Free—bound radiation, 37
Free—free radiation, 37
"freezing," magnetic field, 46
Flute instability, 98

Gas discharge, 24
Glow discharge, 25

Gradient drift, 79
Group velocity, 106
Guiding center, 75

Hartmann number, 59
Hybrid frequencies, 118
Hydrodynamic approximation, 89
Hydromagnetic instabilities, 91

Impact parameter, 145
Impurity, 5
Independent particle model, 66
Ion—acoustic oscillation, 111
Ion plasma frequency, 64
Ion plasma oscillation, 112
Ionization, 35, 141
Ionization energy, 35
Ionization, degree of, 41
Ionosphere, 123
Interchange (flute) instability, 98

Joule heat, 107

Kink instability, 96

Laminar flow, 57
Landau damping, 110, 130
Langmuir probe, 12
Linear oscillation, 105
Lorentz force, 66
Loss cone, 77
Lower hybrid frequency, 118

Magnetic mirror, 5, 76
Magnetic moment, 74
Magnetic pressure, 46
Magnetic probe, 16
Magnetic Reynolds number, 58
Magnetic viscosity, 58
Magnetoacoustic wave, 114
Magnetohydrodynamics, 45
Magnetoplasma, 2, 64, 151
Magnetosonic wave, 118
Maxwellian distribution, 126
Mean free path, 139
Microwave cutoff, 121

Microwave diagnostics, 121
Microwave interferometer, 124
Molecular band, 15
Molecular ion, 35
Multiply-charged ion, 15

Negative ion, 36

Oblique wave, 119
Opacity, plasma, 38
Optical pyrometry, 15
Optical spectroscopy, 15
Optical thickness, 38
Ordinary wave, 117
Oscillatory instability, 130

Pauli principle, 33
Phase velocity, 106
Pinch, 52
Plasma (defined), 22
Plasma frequency, 20, 63, 109
Plasma oscillation, 20, 62, 104
Plasma resonator, 126
Polarization, plasma, 22
Polarization drift, 83
Positive column, 25

Quasi neutrality, 18

Radiation belts, 51
Radiative diffusion, 38
Radiative thermal conductivity, 38
Rail accelerator, 56
Ramsauer effect, 144
Random walk, 138
Recombination, 141
  radiative, 35
  three-body, 35
Relaxation time, 29, 149
Reynolds number, 58
Rogowsky loop, 17
Runaway electron, 146
Running electron number, 128

Saha equation, 41
Saturation current, 12

Sausage instability, 96
Self-sustaining discharge, 25
Shock wave, 134
    collisionless, 135
Simon diffusion, 153
Skin depth, 49
Solar wind, 51
Spectrum,
    discrete, 36
    continuous, 37
Standing wave, 107
Stellarator, 54, 98
Stuart number, 59
Synchrotron radiation, 39

Temperature,
    plasma, 26
    longitudinal, 28
    transverse, 28
Thermodynamic equilibrium, 40
Thermal ionization, 8

Toroidal pinch, 54
Toroidal trap, 53
Transport phenomena, 136
    in magnetic field, 150
Trapped radiation, 38
Traveling wave, 107
Turbulent flow, 57
Two-fluid model, 60

Upper hybrid frequency, 118

Viscous dissipation, 108
Volt−ampere characteristic, 12

Wave propagation, in plasmas, 120
Wave number, 106
Wave vector, 106
Whistler, 126

Zeeman splitting, 17